Power Query Cookbook

Use effective and powerful queries in Power BI Desktop and Dataflows to prepare and transform your data

Andrea Janicijevic

BIRMINGHAM—MUMBAI

Power Query Cookbook

Copyright © 2021 Packt Publishing

Group Product Manager: Kunal Parikh

Publishing Product Manager: Ali Abidi

Senior Editor: Roshan Kumar

Content Development Editor: Tazeen Shaikh

Technical Editor: Rahul Limbachiya

Copy Editor: Safis Editing

Project Coordinator: Aparna Ravikumar Nair

Proofreader: Safis Editing

Indexer: Subalakshmi Govindhan

Production Designer: Prashant Ghare

First published: October 2021

Production reference: 1070921

Published by Packt Publishing Ltd.
Livery Place
35 Livery Street
Birmingham
B3 2PB, UK.

ISBN 978-1-80056-948-5

www.packt.com

To my family, who have always supported me during my studies and my professional experience and have been key in me becoming the woman I am today. To my colleague, Patrik Borosch, who introduced me to this opportunity, and to my manager, Zoran Draganic, who gave me the freedom to pursue this exciting journey.

– Andrea Janicijevic

Contributors

About the author

Andrea Janicijevic is a cloud solution architect and works in the world of analytics and business intelligence, constantly expanding her knowledge in the field of data. From the outset, she has been working on analytics platforms, helping clients to better adopt cloud technology across a wide range of industries and company sizes.

She studied economics and management of innovation and technology at Bocconi University in Milan and during her studies, she started working at Microsoft in 2018. She began working with the Microsoft analytics platform, including Power BI, becoming a trusted technical advisor for business and technical users. She later started collaborating with Packt, accepting the challenge of sharing her experience with Power Query.

About the reviewers

Patrik Borosch is a cloud solution architect for data and AI at Microsoft Switzerland GmbH. He has more than 25 years of BI and analytics development, engineering, and architecture experience and is a Microsoft Certified Data Engineer and a Microsoft Certified AI Engineer. Patrik has worked on numerous significant international data warehouse, data integration, and big data projects. Through this, he has built and extended his experience in all facets, from requirements engineering to data modeling and ETL, all the way to reporting and dashboarding. At Microsoft Switzerland, he supports customers in their journey into the analytical world of the Azure Cloud.

Michiel Rozema is one of Europe's top Power BI experts, living in the Netherlands. He holds a master's degree in mathematics and has worked in the IT industry for over 25 years as a consultant and manager. Michiel was the data insight lead at Microsoft Netherlands for 8 years, during which time he launched Power BI in the country. He is the author of two Dutch books on Power Pivot and Power BI, and is the author of the *Extreme DAX* title with Packt Publishing. Michiel is one of the founders of the Dutch Power BI user group and the initiator of the Power BI Summer School, and has been a speaker at many conferences on Power BI. He has been awarded the Microsoft MVP award since 2019 and, together with fellow MVP Henk Vlootman, runs the consultancy firm Quanto, specializing in Power BI.

Table of Contents

3

Data Exploration in Power Query

4

Reshaping Your Data

5
Combining Queries for Efficiency

6
Optimizing Power Query Performance

7
Leveraging the M Language

8

Adding Value to Your Data

9

Performance Tuning with Power BI Dataflows

10
Implementing Query Diagnostics

Other Books You May Enjoy

Index

Preface

Power Query is a data preparation tool that enables data engineers and business users to connect, reshape, enrich, and transform their data to facilitate relevant business insights and analysis. With Power Query's wide range of features, you can perform no-code transformations and complex M code functions at the same time to get the most out of your data.

This Power Query book will help you to smartly connect to data sources, achieve intuitive transformations, and get to grips with preparation practices. Starting with a general overview of Power Query and what it can do, the book advances to cover more complex topics such as M code and performance optimization. You'll learn how to extend these capabilities by gradually stepping away from the Power Query GUI and into the M programming language. Additionally, the book also shows you how to use Power Query Online within Power BI Dataflows.

By the end of the book, you'll be able to leverage your source data, understand your data better, and enrich it with a full stack of no-code and custom features that you'll learn to design by yourself for your business requirements.

Who this book is for

This book is for data analysts, BI developers, data engineers, and anyone looking for a desk reference guide to learn how Power Query can be used with different Microsoft products to handle data of varying complexity. Beginner-level knowledge of Power BI and the M language will help you to get the most out of this book.

What this book covers

Chapter 1, Getting Started with Power Query, focuses on what Power Query is, how the tool has evolved, and where you can find/use it across Microsoft platforms. Then, we share when to use Power Query within each Microsoft service (Power BI, Excel, Analysis Services, Power Apps, and Azure Data Factory), giving you an idea of how different types of users can leverage the same tool for different purposes.

Chapter 2, Connecting to Fetch Data, shows an overview of connectors. Some best practices will be shared on how to connect to some of the most common connector types. The main ones identified are connections to files, folders, databases, and websites.

Chapter 3, Data Exploration in Power Query, focuses on data exploration features in Power Query. You will learn how to choose a subset of data and explore data profiling tools and query dependencies in order to see at a glance what data you will be dealing with. You will see how to smartly use query and step panes with some shortcuts and examples. Moreover, the schema and diagram views will be explained.

Chapter 4, Reshaping Your Data, focuses on how users can reshape their data. Most common transformation tasks will be shown, sharing best practices that you can apply to a wide range of dataset types. Other than data manipulation and wrangling, some artificial intelligence features such as Cognitive Services will be shown.

Chapter 5, Combining Queries for Efficiency, describes how users can combine different queries. Merge and append possibilities will be explained. Best practices for multiple file combinations are also shown.

Chapter 6, Optimizing Power Query Performance, aims to clarify what features you can leverage to optimize Power Query queries. The setup of parameters and their use will be explained and we will take a deep dive into how to best approach query folding. It is important for you to understand how query folding works and how to apply it in an incremental refresh scenario.

Chapter 7, Leveraging the M Language, gives an outline of M coding. The differences from the DAX language will be clarified and knowledge about how to deal with existing and new queries using M code will be shared. The chapter will focus both on simpler and more advanced scenarios involving M code.

Chapter 8, Adding Value to Your Data, aims to teach you how you can enrich your data by using add column features that range from simpler ones such as columns from examples to more advanced ones such as custom columns. Custom functions will be explored and examples will be shared. The cluster values feature, one of the most recent, will be given as an example.

Chapter 9, Performance Tuning with Power BI Dataflows, explains Power BI dataflows. It will focus on how users can leverage the Power Query engine to create dataflows, how to schedule a refresh, and how to allow other users to build data models by using dataflows as central data sources. The aim is to clarify what the best practices are, such as to prefer dataflows over Power Query in other Microsoft tools, and which are the most common scenarios for their use.

Chapter 10, Implementing Query Diagnostics, focuses on Power Query diagnostics. There is a specific tool for that and it is useful to describe how to use it and interpret its output.

To get the most out of this book

You will need a version of Power BI Desktop installed – the latest, if possible. All code examples have been tested using Power BI Desktop (March 2021 version). They will also work with future version releases too. The Power BI Gateway version used is February 2021 (3000.72.6). Also, future releases work the same way for the features explored in this book (Note: the user interface may differ depending on the version you use or due to future updates).

Software/hardware covered in the book	Operating system requirements
Power BI Desktop	Windows
Power BI Gateway	Windows

The majority of the tasks will need you to install Power BI Desktop, which is a free tool, whereas some of them will require you to have a Power BI Pro license to access the Power BI service on the web. In some recipes, you will need to have a Power BI Premium capacity to access some advanced features, such as Power BI Dataflows. The licenses needed for each chapter are the following:

- *Chapter 1, Getting Started with Power Query*, requires a Power BI Desktop free license.

- *Chapter 2, Connecting to Fetch Data*, requires a Power BI Desktop free license.

- *Chapter 3, Data Exploration in Power Query*, requires a Power BI Desktop free license.

- *Chapter 4, Reshaping Your Data*, requires a Power BI Desktop free license, Power BI Pro, and Power BI Premium.

- *Chapter 5, Combining Queries for Efficiency*, requires a Power BI Desktop free license.

- *Chapter 6, Optimizing Power Query Performance*, requires a Power BI Desktop free license.

- *Chapter 7, Leveraging the M Language*, requires a Power BI Desktop free license.

- *Chapter 8, Adding Value to Your Data*, requires a Power BI Desktop free license, Power BI Pro, and Power BI Premium.

- *Chapter 9, Performance Tuning with Power BI Dataflows*, requires a Power BI Desktop free license, Power BI Pro, and Power BI Premium.

- *Chapter 10, Implementing Query Diagnostics*, requires a Power BI Desktop free license.

This book will help both users who already have some experience with Power Query to discover features they were not aware of and beginner users to approach this tool with concrete examples, leveraging the power of learning by doing. After reading this book, you will have the skills to understand Power Query and keep an eye on `https://powerquery.microsoft.com/en-us/blog/` and `https://powerbi.microsoft.com/en-us/blog/` to enhance the knowledge you have acquired.

If you are using the digital version of this book, we advise you to type the code yourself or access the code from the book's GitHub repository (a link is available in the next section). Doing so will help you avoid any potential errors related to the copying and pasting of code.

Download the example code files

You can download the example code files for this book from GitHub at `https://github.com/PacktPublishing/Power-Query-Cookbook`. If there's an update to the code, it will be updated in the GitHub repository.

We also have other code bundles from our rich catalog of books and videos available at `https://github.com/PacktPublishing/`. Check them out!

Download the color images

We also provide a PDF file that has color images of the screenshots and diagrams used in this book. You can download it here: `https://static.packt-cdn.com/downloads/9781800569485_Colorimages.pdf`.

Conventions used

There are a number of text conventions used throughout this book.

`Code in text`: Indicates code words in text, database table names, folder names, filenames, file extensions, pathnames, dummy URLs, user input, and Twitter handles. Here is an example: "In this recipe, you need to download the `FactResellerSales` CSV file."

A block of code is set as follows:

```
(OldSalesAmount as number, Discount as number, TotalCosts as
number) =>
let
   NetSales = OldSalesAmount - (OldSalesAmount * Discount ) -
TotalCosts
in
    NetSales
```

Bold: Indicates a new term, an important word, or words that you see on screen. For instance, words in menus or dialog boxes appear in **bold**. Here is an example: "Click on **New Source** and select **Text/CSV**."

> **Tips or important notes**
> Appear like this.

Sections

In this book, you will find several headings that appear frequently (*Getting ready*, *How to do it...*, *How it works...*, *There's more...*, and *See also*).

To give clear instructions on how to complete a recipe, use these sections as follows:

Getting ready

This section tells you what to expect in the recipe and describes how to set up any software or any preliminary settings required for the recipe.

How to do it...

This section contains the steps required to follow the recipe.

How it works...

This section usually consists of a detailed explanation of what happened in the previous section.

There's more...

This section consists of additional information about the recipe in order to make you more knowledgeable about the recipe.

See also

This section provides helpful links to other useful information for the recipe.

Get in touch

Feedback from our readers is always welcome.

General feedback: If you have questions about any aspect of this book, email us at customercare@packtpub.com and mention the book title in the subject of your message.

Errata: Although we have taken every care to ensure the accuracy of our content, mistakes do happen. If you have found a mistake in this book, we would be grateful if you would report this to us. Please visit www.packtpub.com/support/errata and fill in the form.

Piracy: If you come across any illegal copies of our works in any form on the internet, we would be grateful if you would provide us with the location address or website name. Please contact us at copyright@packt.com with a link to the material.

If you are interested in becoming an author: If there is a topic that you have expertise in and you are interested in either writing or contributing to a book, please visit authors.packtpub.com.

Share Your Thoughts

Once you've read *Power Query Cookbook*, we'd love to hear your thoughts! Scan the QR code below to go straight to the Amazon review page for this book and share your feedback.

https://packt.link/r/1-800-56948-3

Your review is important to us and the tech community and will help us make sure we're delivering excellent quality content.

1
Getting Started with Power Query

Power Query is a data preparation tool that enables data engineers and business users to connect, reshape, enrich, and transform their data. This allows them to facilitate relevant business insights analysis. Power Query is a technology that strengthens self-service business intelligence with an intuitive and consistent experience. It consists of a graphical interface that facilitates the connection to data sources and the application of different ranges of transformation.

Power Query is not a standalone tool; it can be used inside different tools in two different versions: **Power Query Desktop** and **Power Query Online**. The first version is available in Excel, Power BI, and SQL Server Analysis Services, while the second is available in the Power BI service, Power Apps, Power Automate, Azure Data Factory, Azure Synapse, and Dynamics 365 Customer Insights. Depending on where Power Query is used, users will be able to store reshaped data in different ways: publish datasets to the Power BI service, load data in Azure Data Lake with Common Data Model formatting, and load transformed data to the Dataverse.

The following recipes will be covered in this chapter:

- Installing a Power BI gateway

- Authentication to data sources

- Main challenges that Power Query solves

Technical requirements

In this chapter, you will be using the following:

- **Power BI Desktop**: `https://www.microsoft.com/en-us/download/details.aspx?id=58494`

- **Power BI Pro License**: `https://powerbi.microsoft.com/en-us/power-bi-pro/`

- **Power BI gateway**: `https://powerbi.microsoft.com/en-us/gateway/`

The minimum requirements for installation are as follows:

- **.NET Framework 4.6** (Gateway release August 2019 and earlier)

- **.NET Framework 4.7.2** (Gateway release September 2019 and later)

- A 64-bit version of Windows 8 or a 64-bit version of Windows Server 2012 R2 with current TLS 1.2 and cipher suites

- 4 GB of disk space for performance monitoring logs

You can find the data resources referred to in this chapter at `https://github.com/PacktPublishing/Power-Query-Cookbook/tree/main/Chapter01`.

Installing a Power BI gateway

Power BI users often need to work with data from on-premises sources, such as filesystems, local files available on a PC, and databases not running on the cloud. In order to make this data securely available once the report is published to the web, a Power BI gateway needs to be installed. Microsoft offers two different types of gateway for different scenarios, and their setup can be customized according to specific enterprise configuration requirements, such as proxy, service account, communication, and high availability settings. Users can choose one of the following two types:

- **Standard (or enterprise) mode**: This mode can be used to connect data sources to Power Platform services, Logic Apps, and Analysis Services by multiple users. It has to be run by users with admin rights and is meant for enterprise scenarios.

- **Personal mode**: This mode can be used by single users without the possibility of sharing the files. This version is available for Power BI only. If you want to quickly connect to an Excel file on your local machine and run tests without needing admin rights, this mode is for you. It is meant for testing purposes.

Customers need monitoring options and analysis to decide whether to scale up or scale down the gateway server to improve data movement performance. This recipe aims to help users to decide which type of gateway to install, and to assist with the configuration and monitoring options.

Getting ready

You can refer to this link to download a Power BI gateway: `https://powerbi.microsoft.com/en-us/gateway/`.

In this recipe, we are going to install the standard (enterprise) mode gateway on a local machine. It is recommended, though, to install the gateway on a server, especially in enterprise scenarios.

In this chapter, Power BI Desktop needs to be installed on a machine that has access to the data sources. Access to the Power BI service is also needed.

Download the data files on your local machine.

How to do it...

Once you have downloaded the Power BI gateway, you are ready to start the setup:

1. Define the default path for your gateway resources, accept the terms, and run the installation. Revise the minimum requirements for the machine where the setup is going to be done:

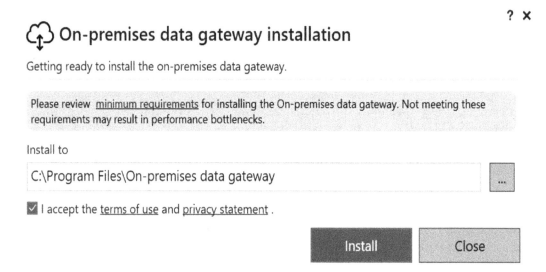

Figure 1.1 – On-premises data gateway installation

2. Enter your work or school Microsoft Office 365 account. This account has to be in the Azure Active Directory tenant, the one shared with Power BI. By entering your organizational account, you will be able to manage gateways and add multiple data sources using the Power BI service portal:

Figure 1.2 – On-premises gateway email admin

3. Click on **Register a new gateway** on this computer:

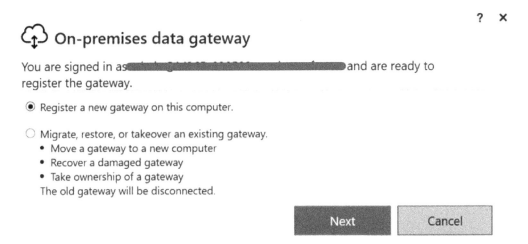

Figure 1.3 – Register a new gateway

4. Give a name to the gateway and create a **Recovery key**. This key is needed if you want to create a gateway cluster (a group of gateways), to migrate your existing gateway, or to take over the gateway's ownership. Once you set the key, click on **Configure**:

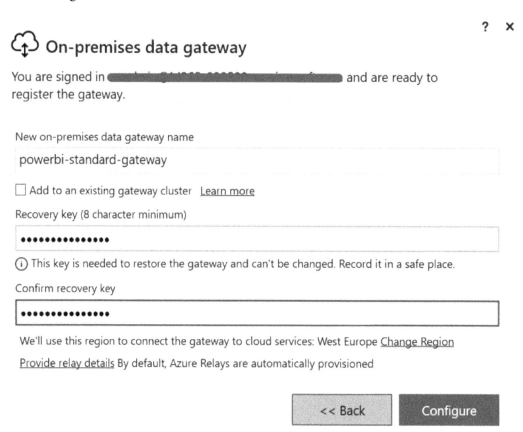

Figure 1.4 – Gateway recovery key definition

At this step, you can decide to change the default region to connect the gateway to cloud services. The default region is the one where your **Power BI** or **Microsoft O365** tenant is located. If you want to change it, you'll select an Azure region, but make sure that the region is close to you.

5. Once the configuration is completed, you should end up with the following view:

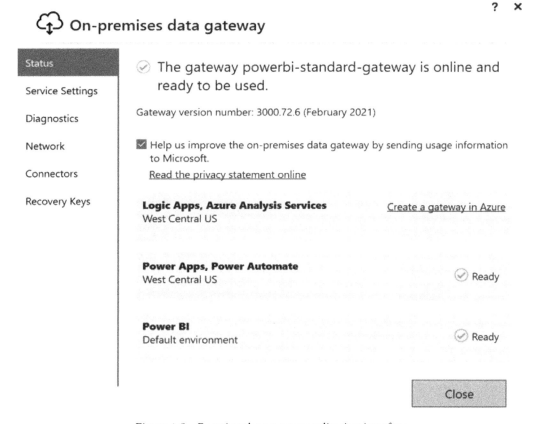

Figure 1.5 – Running the gateway application interface

This will be the default view when you open the gateway application. You can monitor the health of the gateway and see what services can use the same gateway. This application allows the gateway admin to customize the configuration.

Managing the data gateway on the Power BI portal

Once you have installed the gateway and it is running on the machine, you can access `https://powerbi.com` and log in to the Power BI portal with the credentials you use to sign in to the gateway application. When you are logged in, complete the following steps to see how to manage the data gateway:

1. Go to **Manage gateways** and access the section where you can find the running gateway you set up before:

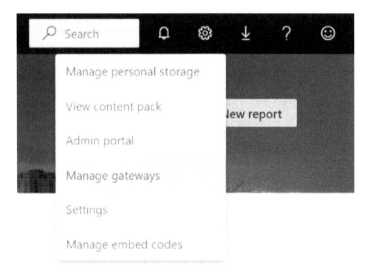

Figure 1.6 – Power BI service: Settings section

2. You will now be able to see gateways settings and administrators:

    a) **Administrators**: If you installed the gateway, you will be an admin by default. You can use these sections to add other administrators:

Figure 1.7 – Managing gateways: adding administrators

b) **Gateway Cluster Settings**: On the left side, you can see the gateway that you configured on your Windows machine. In the following screenshot, you can find a list of data sources that you registered. Click on the three dots and then click on **ADD DATA SOURCE**:

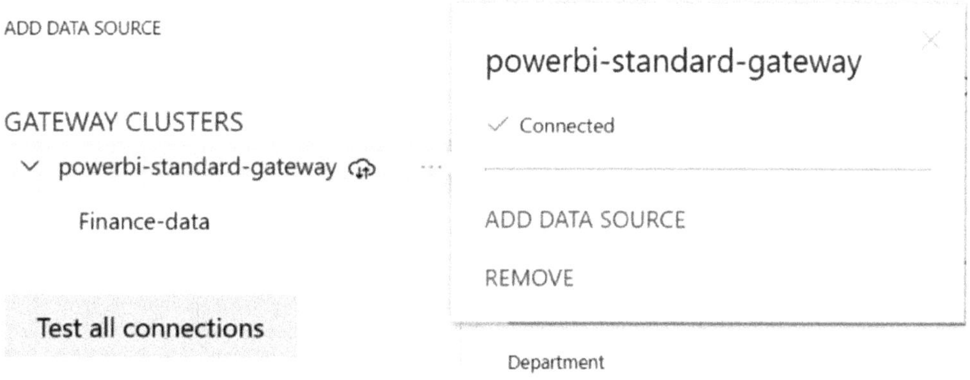

ADD DATA SOURCE

GATEWAY CLUSTERS

∨ powerbi-standard-gateway

Finance-data

Test all connections

powerbi-standard-gateway   ╳

✓ Connected

ADD DATA SOURCE

REMOVE

Department

Figure 1.8 – Managing gateways: cluster view

Name the data source and select the **File** type (you can explore and add other types, such as SQL servers, folders, SAP, and ODBC). You can add on-premises data sources that can be accessed by the gateway installed on a server that belongs to the same data source domain:

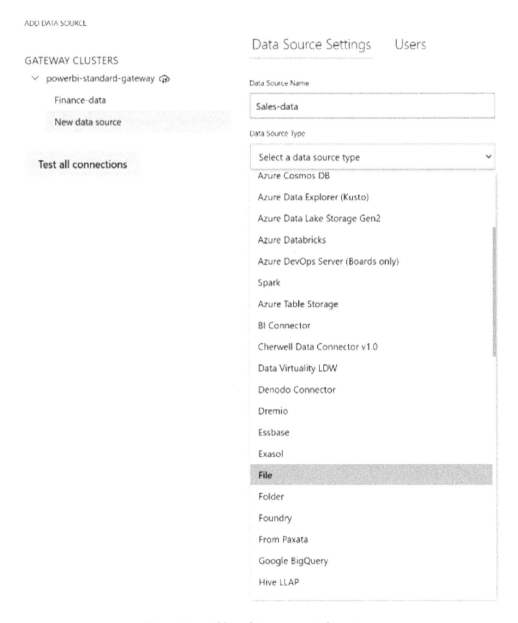

Figure 1.9 – Adding data sources to the gateway

3. You will add one of the files that you downloaded. Add the **Full path** and enter the credentials to access the data source:

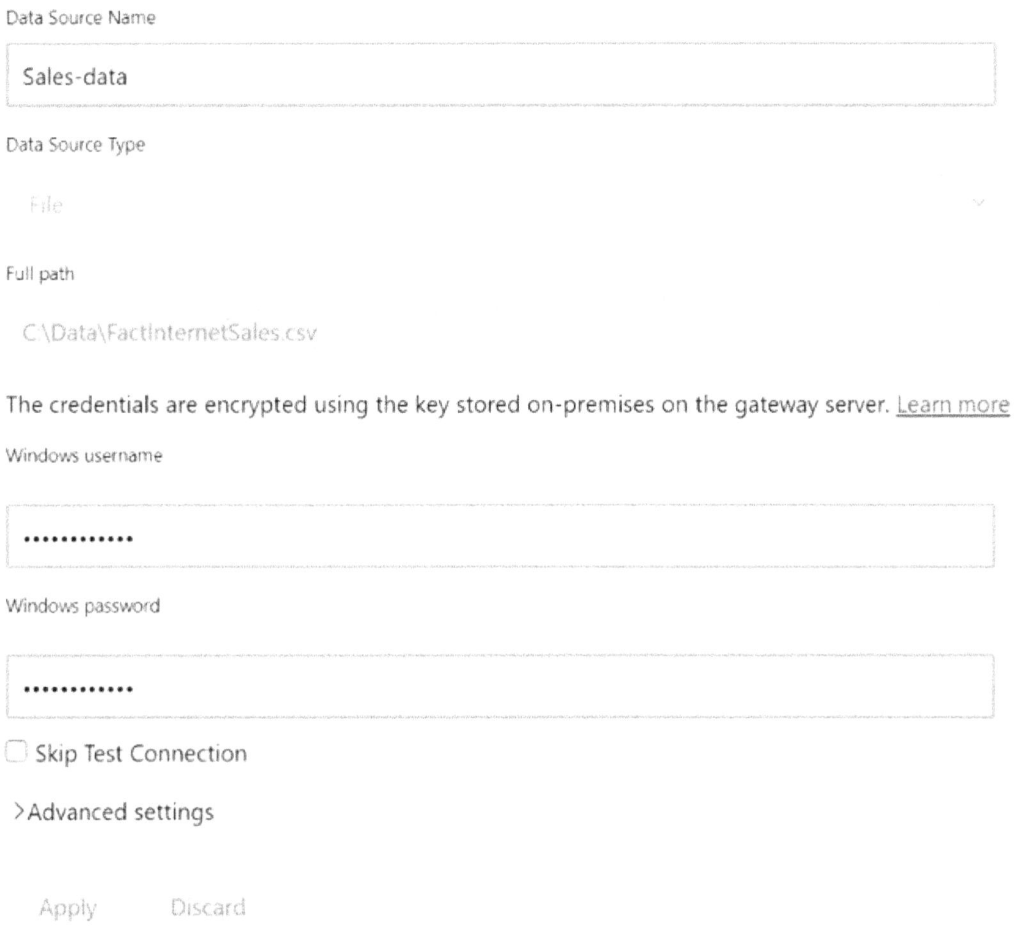

Data Source Name
Sales-data
Data Source Type
File
Full path
C:\Data\FactInternetSales.csv
The credentials are encrypted using the key stored on-premises on the gateway server. Learn more
Windows username
Windows password
Skip Test Connection
>Advanced settings
Apply
Discard

Figure 1.10 – Adding credentials for data sources

4. When you click on **Apply**, Power BI will check the connection and, if the connection succeeds, you will see a **Connection Successful** status:

Data Source Settings
Users
Connection Successful

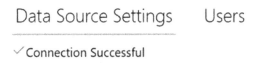

Figure 1.11 – Data source connection outcome

Once you have added data sources to be accessed through the gateway, you have three checkboxes at the end of the page that you can flag:

a) **Allow user's cloud data sources to refresh through this gateway cluster**: Check this box if you plan to perform append and merge operations between on-prem and cloud sources.

b) **Allow user's custom data connectors to refresh through this gateway cluster**: If you have a custom-developed connector built to use the gateway to access data sources, you need to flag this box to refresh data.

c) **Distribute requests across all active gateways in this cluster**: If you have other gateways configured within the same cluster, you can enable this feature to perform load balancing and distribute requests across all active clusters:

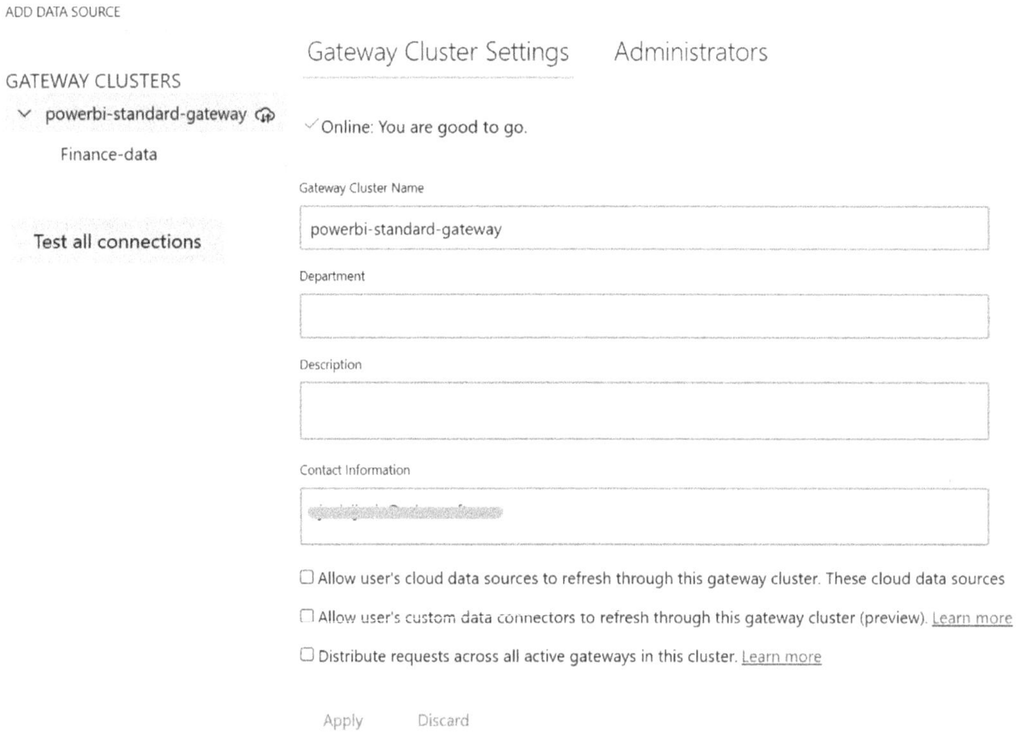

ADD DATA SOURCE

GATEWAY CLUSTERS

∨ powerbi-standard-gateway

Finance-data

Test all connections

Gateway Cluster Settings   Administrators

✓ Online: You are good to go.

Gateway Cluster Name

powerbi-standard-gateway

Department

Description

Contact Information

☐ Allow user's cloud data sources to refresh through this gateway cluster. These cloud data sources

☐ Allow user's custom data connectors to refresh through this gateway cluster (preview). Learn more

☐ Distribute requests across all active gateways in this cluster. Learn more

Apply   Discard

Figure 1.12 – Gateway Cluster Settings

5.   Open Power BI Desktop, connect to one of the data sources you downloaded and that you added in the previous steps, create a report, and publish it on the Power BI service.

Follow these steps to check that your data source is used correctly with the gateway:

1. Once you have published the report and dataset in your workspace, click on the three dots next to the dataset and go to **Settings**:

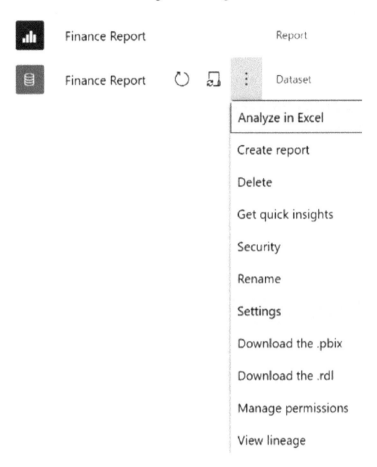

Figure 1.13 – Settings section for datasets

2. Expand the **Gateway connection** section and check that the gateway is running successfully:

Settings for Finance Report

Refresh history

▶ Dataset description

◢ Gateway connection

To use a data gateway, make sure the computer is online and the data source is added in Manage Gateways. If you're using an On-premises data gateway (standard mode), please select the corresponding data sources and then click apply.

Use an On-premises or VNet data gateway

⬤ On

	Gateway	Department	Contact information	Status	Actions
○	☁ powerbi-standard...		⌀⌀⌀⌀⌀⌀⌀⌀...	⊘ Running on⬛⬛⬛⬛, ⬛⬛⬛⬛ select all datasources to use	⚙ ▼

Data sources included in this dataset:

⊘ File{"path":"c:\\data\\finance datasources.xlsx"}   Maps to: [⌄]

Figure 1.14 – Defining dataset sources

3. You can map the data sources included in the dataset to sources that you created in the **Gateway Management** view. Map to **Finance-data** and click on **Apply**. You could also do it the other way around: create a report, publish it, access the following view, click on **Add to gateway**, and be redirected to the **Gateway Management** section, where you can add data sources to your gateway:

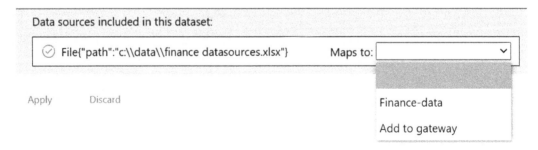

Figure 1.15 – Mapping a dataset to an existing gateway source

How it works

From now on, you can refresh your datasets manually or at scheduled times. The high flexibility is given by the fact that you can create single gateways or create clusters of gateways following this recipe in order to meet high availability and load balancing requirements. For each gateway or cluster of gateways, you can add or remove data sources that use those gateways to create a connection between on-premises environments and cloud services by allowing the following:

- A fast and secure connection
- A gateway that decrypts and uses stored credentials to access on-premises data sources handling authentication
- Easy-to-use and straightforward administration, management, and troubleshooting tools

Authentication to data sources

Power Query provides different connector types to a wide range of data sources. These will be widely explained in the following chapter, but in this recipe, we will concentrate on how authentication works for data sources.

Each connector provides different kinds of authentication. There are six main types of authentication:

- Anonymous
- Windows
- Basic
- Organizational account
- Microsoft account
- Database

Depending on the type of connector and the Power Query version used, different combinations of these options will be available for the user.

Once you connect to your data source and you perform Power Query transformations, you will be able to change the type of authentication and the data source without losing the work you have done.

Getting ready

In this recipe, in order to test different types of connections, you need to have the following data sources to which you can connect:

- An **Azure SQL Database** with **AdventureWorks** data, database credentials, and access through Azure Active Directory Authentication (log in with your Microsoft account)

- A *Parquet file* named `FactInternetSales` in a local folder on your PC

How to do it...

Once you have opened your Power BI Desktop application, you are ready to follow these steps:

1. Go to **Get data** and click on **More**:

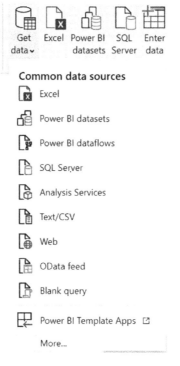

Figure 1.16 – Get data

Search for the **Azure SQL database** connector and select it:

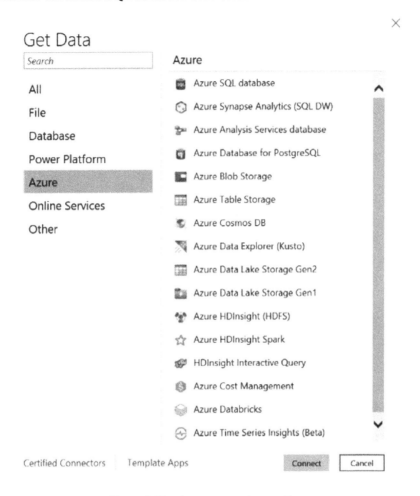

Figure 1.17 – Azure connectors section

2.   Enter the name of the server where you have AdventureWorks data:

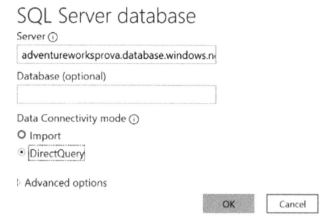

Figure 1.18 – SQL Server database connector

3.   Click on **Database**, enter your database credentials in order to authenticate, and click on **Connect**:

Figure 1.19 – SQL Server authentication

4.   Select the FactInternetSales table when you recreate the AdventureWorks database on your server and click on **Transform Data**:

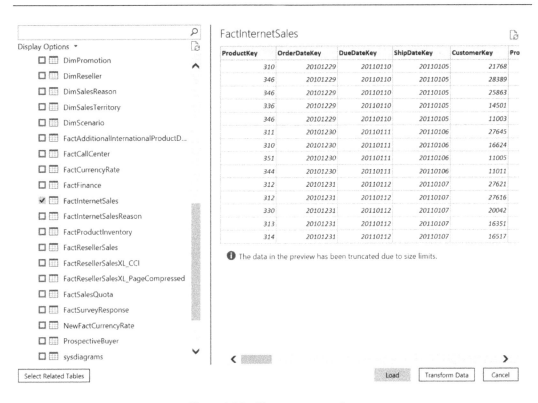

Figure 1.20 – Data source preview

5. A Power Query window pops up, and you are ready to perform some Power Query transformations:

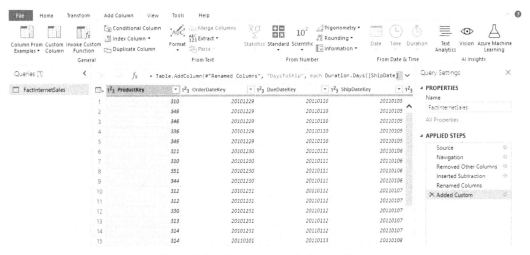

Figure 1.21 – Data preview in Power Query

6. Open your **Advanced Editor** and paste the M code you find in `MCodeChapter1.txt` to reproduce the following Power Query Applied steps:

a) Remove columns

b) Subtraction between two columns

c) Custom column

In the file, you will have to fill in some information regarding your server and the name of your database. Then, all steps will be applied:

Figure 1.22 – Advanced Editor for FactInternetSales

The idea in this recipe is not to focus on the transformations, but to demonstrate that you can change data sources without doing everything from scratch and keeping your M code with your transformations.

Change existing data source permissions

If you want to change permissions, such as switch from database credentials to Microsoft account authentication, you don't have to do everything from scratch, but you can do the following:

1. Click on **Data source settings** and the following window will pop up:

Figure 1.23 – Data source settings

2. From here, you can do the following:

    a) **Change Source**: Click on this to change the name of the server. In this way, you will keep the connector type.

b) **Edit Permissions**: You can change the type of permission used to authenticate. Click on **Edit**, go to the **Microsoft Account** tab, and sign in with your account details:

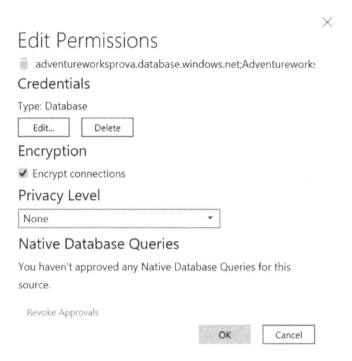

Figure 1.24 – Data source settings: Edit Permissions

You can perform the authentication from the following window:

Figure 1.25 – Data source credentials

c) **Clear Permissions**: You can click on this to delete permissions to connect to this data source. This can be useful when sharing the file with other users and you want to clear permissions. Be careful because once you clear it, you will need to re-connect and re-authenticate.

3. Click on **Save** and close **Data source settings**. You can see that nothing has changed, and you kept all your transformations.

Change connector type

What if you want to keep the same transformations, but change the type of connector? You can do it by using the Advanced Editor.

Complete the following steps to use a Parquet file as a data source. You can find `InternetSales.parquet` in the GitHub folder:

1. From the **Power Query** view, go to **Get Data** and search for the **Parquet** connector. Paste the path where you downloaded this file on your local computer:

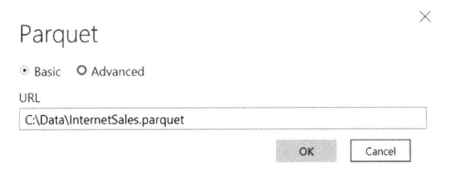

Parquet

○ Basic   O Advanced

URL

C:\Data\InternetSales.parquet

OK   Cancel

Figure 1.26 – Parquet connector

2. Click on **OK** and open **Advanced Editor** on the `InternetSales` query:

InternetSales

```
let
    Source = Parquet.Document(File.Contents("C:\Data\InternetSales.parquet"))
in
    Source
```

Figure 1.27 – Advanced Editor for InternetSales

3. Copy the following line of code and close **Advanced Editor**:

```
Source = Parquet.Document(File.Contents(«C:\Data\
InternetSales.parquet»))
```

4. Select the `FactInternetSales` query and open **Advanced Editor**. Delete the first three lines of code, as shown in the following screenshot:

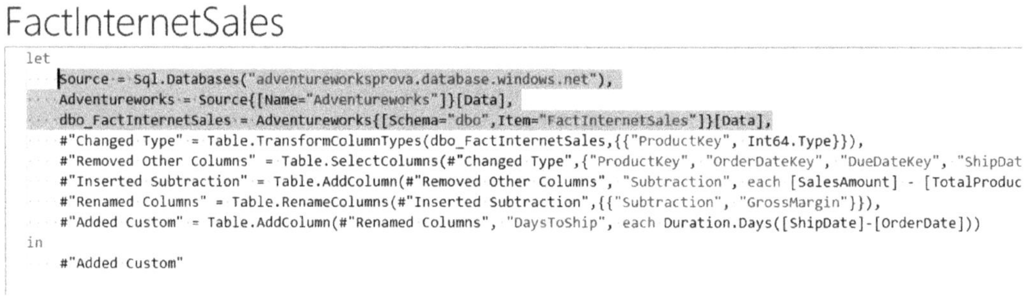

FactInternetSales

```
let
    Source = Sql.Databases("adventureworksprova.database.windows.net"),
    Adventureworks = Source{[Name="Adventureworks"]}[Data],
    dbo_FactInternetSales = Adventureworks{[Schema="dbo",Item="FactInternetSales"]}[Data],
    #"Changed Type" = Table.TransformColumnTypes(dbo_FactInternetSales,{{"ProductKey", Int64.Type}}),
    #"Removed Other Columns" = Table.SelectColumns(#"Changed Type",{"ProductKey", "OrderDateKey", "DueDateKey", "ShipDat
    #"Inserted Subtraction" = Table.AddColumn(#"Removed Other Columns", "Subtraction", each [SalesAmount] - [TotalProduc
    #"Renamed Columns" = Table.RenameColumns(#"Inserted Subtraction",{{"Subtraction", "GrossMargin"}}),
    #"Added Custom" = Table.AddColumn(#"Renamed Columns", "DaysToShip", each Duration.Days([ShipDate]-[OrderDate]))
in
    #"Added Custom"
```

Figure 1.28 – Advanced Editor for FactInternetSales: code selection

Now, paste the code you copied in *Step 3*:

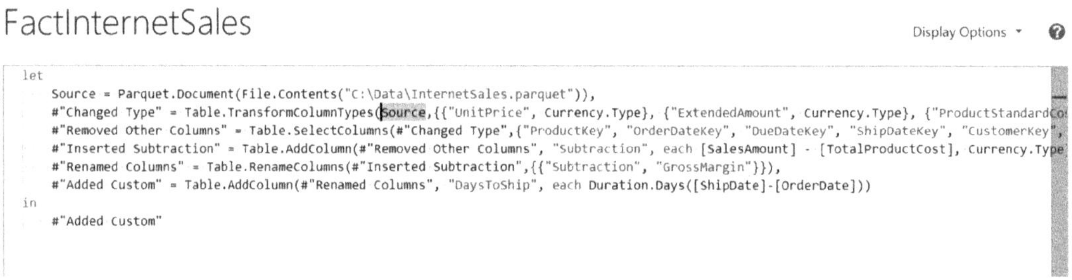

FactInternetSales   Display Options ▼ ❓

```
let
    Source = Parquet.Document(File.Contents("C:\Data\InternetSales.parquet")),
    #"Changed Type" = Table.TransformColumnTypes(Source,{{"UnitPrice", Currency.Type}, {"ExtendedAmount", Currency.Type}, {"ProductStandardCo
    #"Removed Other Columns" = Table.SelectColumns(#"Changed Type",{"ProductKey", "OrderDateKey", "DueDateKey", "ShipDateKey", "CustomerKey"
    #"Inserted Subtraction" = Table.AddColumn(#"Removed Other Columns", "Subtraction", each [SalesAmount] - [TotalProductCost], Currency.Type
    #"Renamed Columns" = Table.RenameColumns(#"Inserted Subtraction",{{"Subtraction", "GrossMargin"}}),
    #"Added Custom" = Table.AddColumn(#"Renamed Columns", "DaysToShip", each Duration.Days([ShipDate]-[OrderDate]))
in
    #"Added Custom"
```

Figure 1.29 – Advanced Editor for FactInternetSales: code replaced

5. At `#"Changed Type"`, replace `dbo_FactInternetSales` with `Source` in order to correctly recall the previous step in Power Query.

6. Close **Advanced Editor** and refresh the query. You will see that, even by changing the data source (from Azure SQL Database to a local Parquet file), the applied steps will be executed.

How it works

Power Query, thanks to its flexibility, offers different options to change data sources and edit data permissions intuitively in order to not waste time managing custom connections when data has to be explored and refreshed. Imagine that you have to deploy to a different environment with a different data source connection: you don't need to rebuild the queries, but just change the source and deploy it to the correct environment.

Main challenges that Power Query solves

By going through the previous recipe, you had the chance to see the variety of transformations and preparation options that Power Query provides.

Power Query aims to solve some of the traditional challenges linked to data analysis:

- Responding to the need for a *low code* tool for business analysts who need to make corrections quickly without doing things from scratch

- Managing *different data types* and *volumes* with quick transformations

- Having a *consistent experience* across platforms and enabling different users to collaborate even if they are using different tools

Getting ready

For this recipe, you need to open the PBIX file provided within the materials. Moreover, you need a *Power BI Premium Capacity* or a *Power BI Embedded capacity* allocated and linked to your Power BI tenant.

How to do it...

Starting from the first challenge, when Power Query was introduced in Excel, it changed the way business users used Excel. It helped users to perform tasks and transformations that would be more complex with a combination of Excel formulas.

Using a low code tool

With a friendly interface, Power Query makes users perform the following actions with a *low code* approach:

- **Connect to data sources** with built-in connectors:

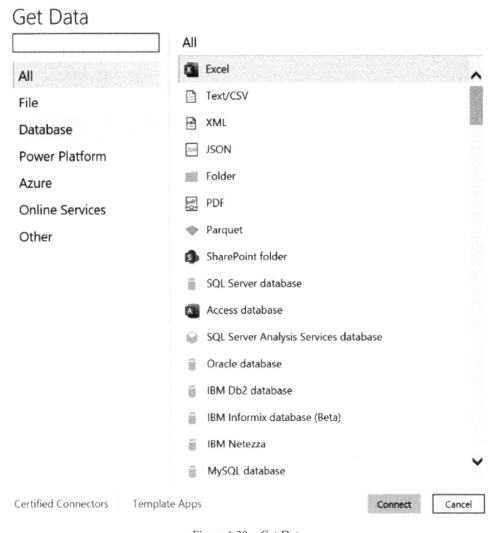

Figure 1.30 – Get Data

- Use a wide set of options and perform a wide range of transformations by clicking a few buttons:

a) Transform data:

Figure 1.31 – Power Query: Transform section

b) Add columns and dimensions, and perform low code or custom operations:

Figure 1.32 – Power Query: Add Column section

c) Have a *persistent* trace of every change done thanks to the **APPLIED STEPS** section:

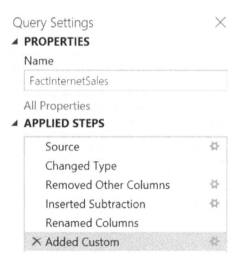

Figure 1.33 – Applied Steps section

d) Once you've performed a single step, you can always go back and change it by clicking on the gear icon:

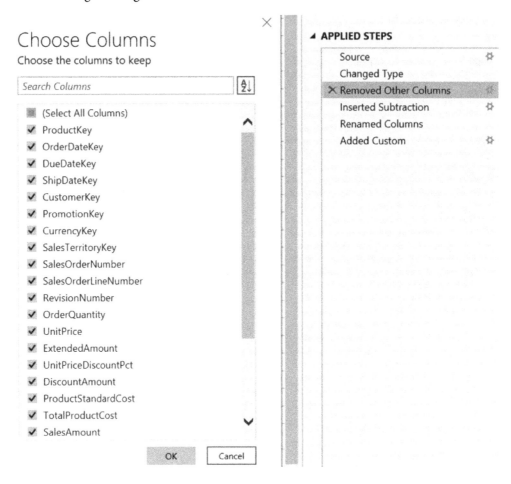

Figure 1.34 – Editing a single step

Managing data sources and data volumes

With the next challenge, users may think that different data sources will be treated and shown differently, but in fact, Power Query standardizes how data can be explored. Once you connect either to SAP or a web page or to Google analytics, once you expand tables, you will be able to perform the same transformations, to aggregate and merge queries as if they come from the same source. All the queries listed here are at the same level.

Often, users do not know how to handle large volumes of complex data in Power Query because they end up with low-performing Power BI files since they encounter a memory restriction because of where they are running Power BI Desktop. It's important to optimize transformations, as shown in the following examples:

- Set the right *data type* for each column, and be careful with the no-datatype column indicated by **ABC123**:

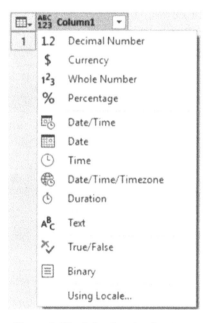

Figure 1.35 – Selecting the data type

- Try to combine the same transformations in fewer steps. Try not to do this:

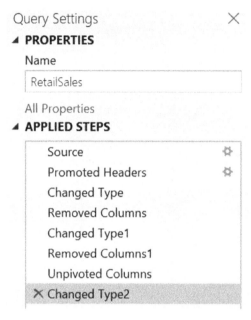

Figure 1.36 – Applied steps: combining transformations

Merge steps to run a unique step for changing the type and another one for removing columns, as in the following example:

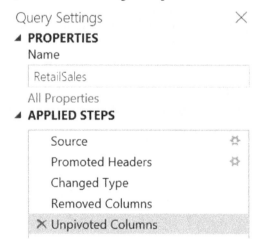

Figure 1.37 – Applied steps: consolidated transformations

You will get the same results but with more compact code and making your transformations more consistent and more easily interpreted by other users.

- Use parameters and range filters to reduce the working dataset for developing your queries. This will lead to a PBIX file size reduction, a reduction in your machine memory usage, and a reduction in the time it takes to load your final model to the cloud. This topic will be covered in *Chapter 6, Optimizing Power Query Performance*.

It is possible to expand tables' volumes once the model has been deployed.

How volume is handled in Power Query depends on which platform is running, because there are different engines underneath: there are differences in running queries on Power Query Online and Desktop.

For example, when running Power Query Online on the Power BI service (a feature that is called Power BI Dataflows) in a Premium Capacity (or, alternatively, Power BI Embedded), you can configure the resources allocated to perform transformations:

- Go to the **Admin portal** when you're logged in to the Power BI service:

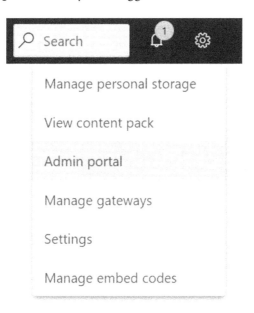

Figure 1.38 – Opening Admin portal in the Power BI service

- Now, navigate to **Capacity settings**:

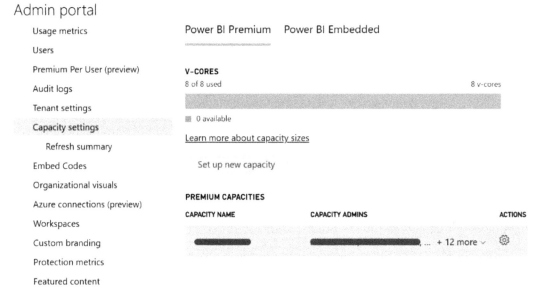

Figure 1.39 – Admin portal in the Power BI service

- Click on **Capacity settings**, expand the **Workloads** section, and navigate until you find **Dataflows**:

Figure 1.40 – Capacity management in the Power BI service

Once you have expanded the **Workloads** section, you can see a part dedicated to dataflows, as shown in the following screenshot:

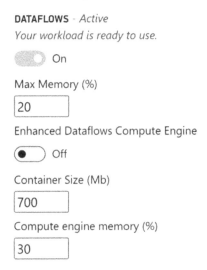

Figure 1.41 – Capacity management: dataflows

- This section allows us to optimize and control Dataflows performance. You can set **Max Memory (%)** relative to the overall capacity node that can be used by dataflows; enable/disable **Enhanced Dataflows Compute Engine**, which provides up to 20x faster performance when working at scale; define the maximum **Container Size** used for each entity (the default value is 700 MB, but can be set up to 5,000 MB); and determine the maximum memory to be used with the compute engine.

Having a consistent experience across platforms

Power Query can be used as a common ground technology between different cloud services. Every tool that contains Power Query is designed to be used in different contexts, teams, and company divisions.

The idea is to enable users with different skills and different tools to collaborate. Different types of users within a company can use the same tool, Power Query, integrated and running seamlessly within the Microsoft platform.

Imagine a business analyst who requires a set of transformations that they developed locally in their Power BI Desktop tool, and they want these steps to be replicated by a data engineer in IT in order to scale this dataflow within the enterprise. The benefit of having Power Query in a service such as Azure Data Factory aims to solve this. Ideally, the business analyst can share their M code with IT, and they can analyze transformations and replicate them with the same language within a Power Query activity in Data Factory.

In this way, Power Query is a tool for agile collaboration in the data environment.

2

Connecting to Fetch Data

One of the main aspects of Power Query is the wide range of data connectors. It offers a varied range of connection options and users can connect to data sources that reside on the cloud, on premises, and in local directories intuitively.

The idea is to treat all data sources at the same level and users (once they select the data they want to transform coming from different sources) can operate and combine them without caring about the data sources' original structure.

In this chapter, there will be an overview of connectors, and we will cover some of the best practices for how to connect to some of the most common connector types.

The recipes that will be covered in this chapter are the following:

- Getting data and connector navigation
- Creating a query from files
- Creating a query from a folder
- Creating a query from a database
- Creating a query from a website

Technical requirements

For this chapter, you will be using the following:

- Power BI Desktop: `https://www.microsoft.com/en-us/download/details.aspx?id=58494`
- A Power BI Pro license: `https://powerbi.microsoft.com/en-us/power-bi-pro/`
- **Minimum requirements** for installation:

  a) .NET Framework 4.6 (Gateway release August 2019 and earlier)

  b) .NET Framework 4.7.2 (Gateway release September 2019 and later)

  c) A 64-bit version of Windows 8 or a 64-bit version of Windows Server 2012 R2 with current TLS 1.2 and cipher suites

  d) 4 GB of disk space for performance monitoring logs

You can find the data resources referred to in this chapter at `https://github.com/PacktPublishing/Power-Query-Cookbook/tree/main/Chapter02`.

Getting data and connector navigation

Power Query, thanks to its interface, offers an easy way to connect to data sources. In the previous chapter, you saw different authentication types, but here you will get an overview of the connector types and learn which one fits best. You will also learn the difference between preview (or beta) and general availability connectors.

Getting ready

For this recipe, you need to have Power BI Desktop running on your machine.

How to do it...

Open Power BI Desktop and you will be ready to perform the following steps:

1. The first step in every version of the **Power Query** tool, whether it is the online or desktop version, is to click on **Get data**:

Figure 2.1 – Get data in Power Query Desktop (left) and Get data in Power Query online (right)

2. Once you expand the **Get data** section, you will end up with the following view in the **Power Query Desktop** version:

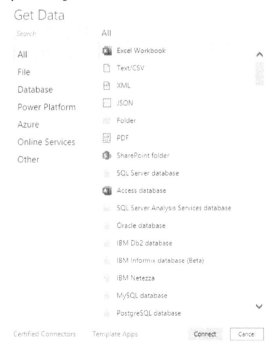

Figure 2.2 – Get Data All connectors view in Power Query Desktop

And if you expand the same section in the **Power Query online** version, you will see the following:

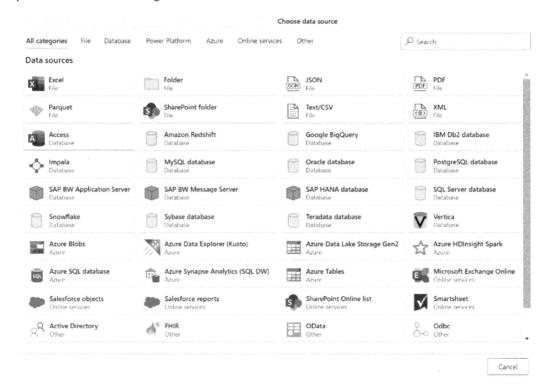

Figure 2.3 – Get Data All categories view in Power Query online

Both versions have the following connectors divided into the same categories:

- **File**: You can connect to different types of files, such as Excel, CSV/TXT, XML, JSON, Folder, PDF, and Parquet.

- **Database**: You can connect to all mainstream databases such as Microsoft, Oracle, IBM, open source databases (MySQL, PostgreSQL, and MariaDB), Teradata, SAP, Amazon Redshift, Google BigQuery, Snowflake, and many others. This wide variety allows the user able to connect to the different sources and not have concerns about having the required data in only one standard data source.

- **Power Platform**: You can connect live to Power BI datasets already published in the Power BI service. You will have the ability to connect to already prepared and transformed queries with the Power BI dataflow connectors and perform additional steps without doing everything from scratch.

- **Azure**: You can connect to all Azure Data Services sources, such as Azure SQL Database, Azure Synapse, Azure Data Lake Storage, and to Azure open source services such as Azure Databricks and Azure HDInsight.

- **Online Services**: You can connect to a wide range of third-party services and use native connectors to the Dynamics platform, Salesforce, Google Analytics, and other services that are continuously updated and released.

- **Other**: This category collects more *generic* connectors, such as web connectors (used for getting data from websites, to make API calls, or to import files from the web), OData feeds, ODBC, and R and Python scripts. This set of connectors allows users to leverage some common connection logic that is used in other tools that can also be replicated with Power Query.

Users have to check what connectors are available in each version of Power Query – either the desktop or online version – and they have to research new connectors' availability. There are new ones both in beta (as shown in the following figure) and a general availability version with every release of Power Query. This list is constantly updated in the Microsoft documentation:

Azure Time Series Insights (Beta)

Figure 2.4 – Connector in the preview example

Creating a query from files

Power Query users (when they start to use and explore the tool) usually start by connecting to a local file. They can see from the start that the main file types are supported and each of these will display data in a readable format.

In this recipe, we will connect to an Excel file and see how to navigate and expand the different sheets and how to connect to cut-off text/CSV files.

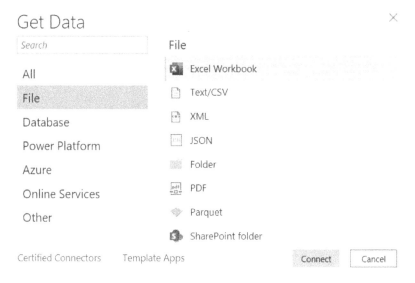

Figure 2.5 – Get Data File section

Getting ready

In this recipe, in order to test different types of file connectors, you need to download the following files in a local folder:

- The AdventureWorksSales Excel file
- The FactResellerSales CSV file

In this example, we will refer to the C:\Data folder.

How to do it...

Once you have opened your Power BI Desktop application, perform the following steps:

1. Go to **Get data** and click on **Excel workbook**:

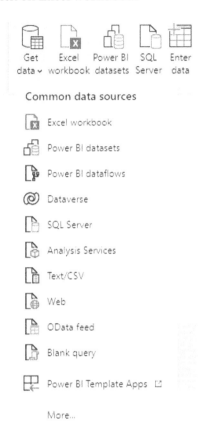

Figure 2.6 – Get data Excel connector

2. Navigate to your local folder where you saved the Excel file, select it, and open it:

Name	Date modified	Type	Size
AdventureWorksSales.xlsx	8/10/2021 7:29 AM	XLSX File	13,988 KB
Finance datasources.xlsx	8/10/2021 7:29 AM	XLSX File	12,175 KB

Figure 2.7 – Local folder view

3.   Once you open it, the following window will pop up:

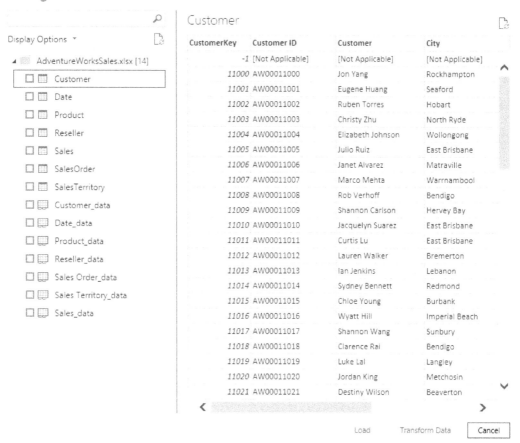

Figure 2.8 – Excel data preview

Each item in the left pane matches an item in the Excel file. By only clicking on an item, you will see a preview of the data in the right pane and if you check it, you will include the item in the Power Query view. Therefore, flag the following queries: **Customer**, **Date**, and **Product**. Click on **Transform Data**.

4.   Each sheet will correspond to a query. From now on, you can perform all transformations as you would with any other data source type:

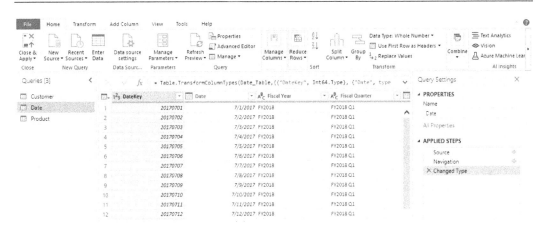

Figure 2.9 – Power Query interface

Let's add a connection to a CSV file:

1. Click on **Get data** and select the **Text/CSV** connector:

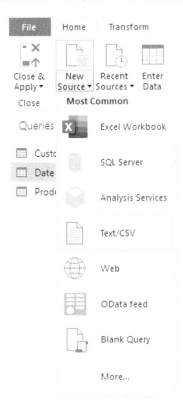

Figure 2.10 – Get data Text/CSV connector

2.   Navigate to the local folder where you saved the `FactResellerSales` CSV file. Select it and open it as in the previous section with the Excel file. The following window will pop up:

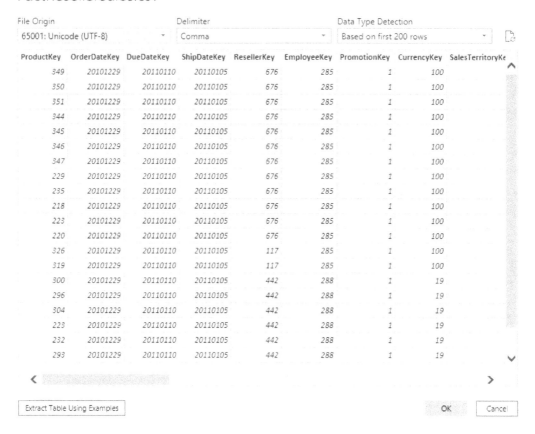

Figure 2.11 – CSV data preview

For each file, you can define the following:

a) **File Origin**: Define the file encoding (in this case, we will keep the default **Unicode UTF-8**).

Figure 2.12 – Define the file encoding

b) **Delimiter**: Select the right delimiter (in this case, we will keep the default **Comma**):

Figure 2.13 – Define the delimiter

c.) **Data Type Detection**: This will refer to the first applied step in Power Query when it detects data types for each column (in this case, we will detect data types based on the first 200 rows):

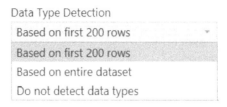

Figure 2.14 – Define Data Type Detection

3.   On the bottom left of this window, you can also extract information from the CSV file by clicking on **Extract Table Using Examples**:

Figure 2.15 – Extract Table Using Examples button

The following section will appear:

Extract Table Using Examples - FactResellerSales.csv

File Origin
65001: Unicode (UTF-8)

File Sample Size
First 200 rows

ProductKey,OrderDateKey,DueDateKey,ShipDateKey,ResellerKey,EmployeeKey,PromotionKey,CurrencyKey,SalesTerritoryKey,SalesOrde
349,20101229,20110110,20110105,676,285,1,100,5,SO43659,1,1,1,"2024,994","2024,994",0,0,"1898,0944","1898,0944","2024,994","
350,20101229,20110110,20110105,676,285,1,100,5,SO43659,2,1,3,"2024,994","6074,982",0,0,"1898,0944","5694,2832","6074,982","
351,20101229,20110110,20110105,676,285,1,100,5,SO43659,3,1,1,"2024,994","2024,994",0,0,"1898,0944","1898,0944","2024,994","
344,20101229,20110110,20110105,676,285,1,100,5,SO43659,4,1,1,"2039,994","2039,994",0,0,"1912,1544","1912,1544","2039,994","
345,20101229,20110110,20110105,676,285,1,100,5,SO43659,5,1,1,"2039,994","2039,994",0,0,"1912,1544","1912,1544","2039,994","
346,20101229,20110110,20110105,676,285,1,100,5,SO43659,6,1,2,"2039,994","4079,988",0,0,"1912,1544","3824,3088","4079,988","

Column1	+
1	
+	

OK   Cancel

Figure 2.16 – Extract Table Using Examples interface

4. You can define your columns and which data to extract by filling in the table at the bottom. Have a look at the following example: name the first column **ResellerKey** and write in the first row the value 676, which is the first **ResellerKey** value you see in the example, and click on **Enter**:

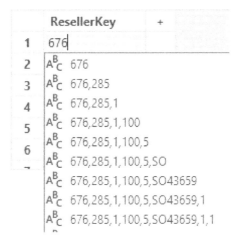

Figure 2.17 – Insert values example

5. If you look at row 5 in *Figure 2.18* (the left image), you can see that a wrong value has been detected. In this case, you can click on it and insert the right one and you will observe how all values in the column will be corrected:

Figure 2.18 – Insert value detail example (left) and fill in missing or wrong values (right)

6. You can add a second column and repeat the steps done with the first. Name the second column **EmployeeKey** and insert the first value. Click **Enter** and you will see the corresponding rows filled:

	ResellerKey	EmployeeKey	+
1	676	285	
2	676	285	
3	676	285	
4	676	285	
5	676	285	
6	676	285	
7	676	285	
8	676	285	
9	676	285	
10	676	285	
11	676	285	
12	676	285	

Figure 2.19 – Create a second column example

7. At the end, click on **OK** and you will see the CSV in the Power Query interface as shown in the following screenshot:

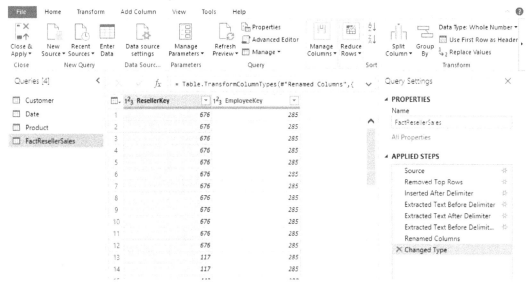

Figure 2.20 – Power Query interface

In the **APPLIED STEPS** section, you will see some activities mapped as a result of **Extract Table Using Examples** performed previously.

How it works...

Power Query, thanks to these file connectors, allows users to connect to single files and perform some pre-transformation tasks allowing them to load just relevant data in the usual interface. However, these connectors – Excel, TXT/CSV, and also Parquet file are related to single files. We will see in the following recipe how to connect to multiple files.

Creating a query from a folder

After playing with single files as the previous recipe showed, you need to load more files as their analytical workloads grow. If files are organized in folders, users can leverage a folder connector to load multiple files. Imagine having a collection of CSV files where each contains sales data for a specific day. What if we want to connect to a folder that contains these files, and we want to load them in Power Query as a single table? The way to go is to leverage the folder connector.

In this recipe, we will see how to connect to a folder with sales data in CSV format and a folder with finance data in Excel format (each file contains multiple sheets).

Getting ready

In this recipe, in order to test different types of file connectors, you need to download the following folders, each containing a set of files:

- The `CSVFiles` folder containing the following CSV files:

InternetSales20110123	Microsoft Excel Com...	2 KB
InternetSales20110124	Microsoft Excel Com...	3 KB
InternetSales20110125	Microsoft Excel Com...	3 KB
InternetSales20110126	Microsoft Excel Com...	2 KB
InternetSales20110127	Microsoft Excel Com...	2 KB
InternetSales20110128	Microsoft Excel Com...	2 KB
InternetSales20110129	Microsoft Excel Com...	2 KB
InternetSales20110130	Microsoft Excel Com...	1 KB
InternetSales20110131	Microsoft Excel Com...	2 KB
InternetSales20110201	Microsoft Excel Com...	2 KB

Figure 2.21 – Local folder with CSV files

- The `ExcelFiles` folder containing the following Excel files:

FinanceData-OnlineChannel	Microsoft Excel Worksheet	9.268 KB
FinanceData-RetailStoreChannel	Microsoft Excel Worksheet	12.201 KB
FinanceData-TelevisionChannel	Microsoft Excel Worksheet	12.203 KB

Figure 2.22 – Local folder with Excel files

In this example, I will refer to the following paths:

a) `C:\Data\ExcelFiles`

b) `C:\Data\CSVFiles`

You can find the folders and the related files referred to in this chapter at `https://github.com/PacktPublishing/Power-Query-Cookbook/tree/main/Chapter02/ExcelFiles` and `https://github.com/PacktPublishing/Power-Query-Cookbook/tree/main/Chapter02/CSVFiles`.

How to do it...

Open the Power BI Desktop application and perform the following steps:

1. Go to **Get data**, click on **Folder**, and the following window will pop up. You can directly enter your folder path or click on **Browse...** and select it from the usual browsing section of your machine:

Figure 2.23 – Folder connector

2. Once you click on **OK**, you will see the following section with a list of files contained in the folder:

C:\Data\CSVFiles

Content	Name	Extension	Date accessed	Date modified	Date created	Attributes	Folder Path
Binary	InternetSales20110123.csv	.csv	8/10/2021 9:26:44 AM	8/10/2021 9:25:17 AM	8/10/2021 9:26:44 AM	Record	C:\Data\CSVFiles\
Binary	InternetSales20110124.csv	.csv	8/10/2021 9:26:44 AM	8/10/2021 9:25:18 AM	8/10/2021 9:26:44 AM	Record	C:\Data\CSVFiles\
Binary	InternetSales20110125.csv	.csv	8/10/2021 9:26:44 AM	8/10/2021 9:25:18 AM	8/10/2021 9:26:44 AM	Record	C:\Data\CSVFiles\
Binary	InternetSales20110126.csv	.csv	8/10/2021 9:26:44 AM	8/10/2021 9:25:17 AM	8/10/2021 9:26:44 AM	Record	C:\Data\CSVFiles\
Binary	InternetSales20110127.csv	.csv	8/10/2021 9:26:44 AM	8/10/2021 9:25:17 AM	8/10/2021 9:26:44 AM	Record	C:\Data\CSVFiles\
Binary	InternetSales20110128.csv	.csv	8/10/2021 9:26:44 AM	8/10/2021 9:25:17 AM	8/10/2021 9:26:44 AM	Record	C:\Data\CSVFiles\
Binary	InternetSales20110129.csv	.csv	8/10/2021 9:26:44 AM	8/10/2021 9:25:17 AM	8/10/2021 9:26:44 AM	Record	C:\Data\CSVFiles\
Binary	InternetSales20110130.csv	.csv	8/10/2021 9:26:44 AM	8/10/2021 9:25:18 AM	8/10/2021 9:26:44 AM	Record	C:\Data\CSVFiles\
Binary	InternetSales20110131.csv	.csv	8/10/2021 9:26:44 AM	8/10/2021 9:25:18 AM	8/10/2021 9:26:44 AM	Record	C:\Data\CSVFiles\
Binary	InternetSales20110201.csv	.csv	8/10/2021 9:26:44 AM	8/10/2021 9:25:17 AM	8/10/2021 9:26:44 AM	Record	C:\Data\CSVFiles\

Figure 2.24 – How files from the folder are displayed

At the bottom right, you can see some actions to perform:

a) **Combine & Transform Data**: You can combine data by appending existing data at this phase and open Power Query.

b) **Combine & Load**: You can append tables, load them, and start creating reports or analyzing data with Excel.

c) **Load**: Load this list into the Power BI dataset as it is.

d) **Transform Data**: This opens the Power Query interface and allows you to do custom transformations.

3. Click on **Transform Data** and you will see the following columns:

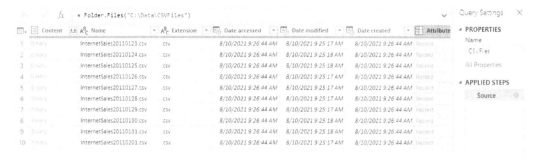

Figure 2.25 – List of files in the Power Query view

From here, you can do one of these actions:

a) Expand a single CSV by clicking on **Binary** in the **Content** column:

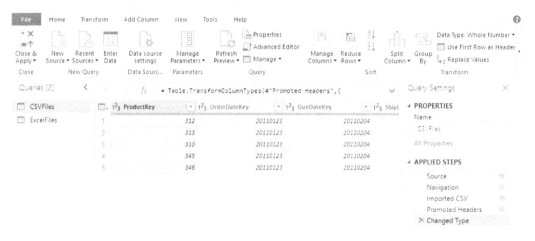

Figure 2.26 – Expanded table

b) Expand the **Attributes** column with some relevant information:

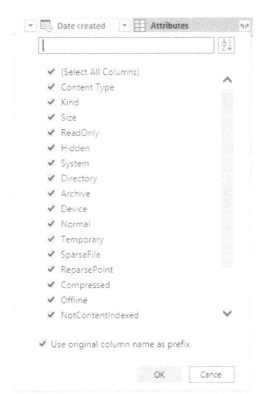

Figure 2.27 – Expand the Attributes column

c) Combine files by clicking on the icon on the right, which means **Combine**:

Figure 2.28 – The Combine icon on the Content column

Data combination is a concept that will be widely explored in *Chapter 5, Combining Queries for Efficiency.*

Now we will repeat the same steps but with the other folder containing Excel files:

1. The view that opens is the following:

Figure 2.29 – List of Excel files in the Power Query interface

It is very similar to the one we saw previously because you can perform the following actions:

a) If you click on **Binary** in row 1, you will end up with a table with a list of the sheets contained in the Excel file `FinanceData-OnlineChannel`. If you click on **Table** in row 1, you will expand the sheet **Sales**:

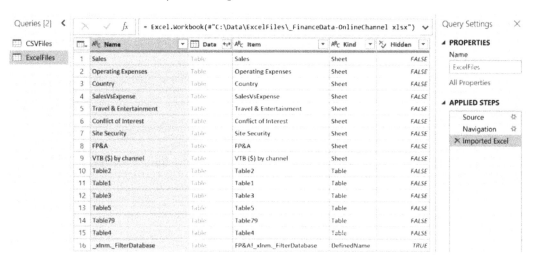

Figure 2.30 – First level of the expanded table

b) If you click on **Combine**, the following window will pop up:

Combine Files

Select the object to be extracted from each file. Learn more

Sample File: First file

	Sales					
	Fiscal Year	Half Yearly	Fiscal Quarter	Fiscal Month	Calendar Month	
Display Options ▾	FY16	H2 Forecast	Q3		8 Feb	
▲ Parameter1 [15]	FY16	H2 Forecast	Q4		12 Jun	
Table1	FY16	H1 Actual	Q1		2 Aug	
Table2	FY16	H2 Forecast	Q3		7 Jan	
Table3	FY16	H1 Actual	Q2		6 Dec	
Table4	FY16	H1 Actual	Q1		1 Jul	
Table5	FY16	H1 Actual	Q2		5 Nov	
Table79	FY16	H1 Actual	Q1		3 Sep	
Conflict of Interest	FY16	H2 Forecast	Q4		10 Apr	
Country	FY16	H1 Actual	Q2		4 Oct	
FP&A	FY16	H2 Forecast	Q3		9 Mar	
Operating Expenses	FY16	H2 Forecast	Q4		11 May	
Sales	FY16	H2 Forecast	Q4		11 May	
SalesVsExpense	FY16	H2 Forecast	Q4		12 Jun	
	FY16	H2 Forecast	Q4		10 Apr	
	FY16	H1 Actual	Q1		1 Jul	
	FY16	H1 Actual	Q1		2 Aug	
	FY16	H2 Forecast	Q4		10 Apr	
	FY16	H1 Actual	Q1		3 Sep	
	FY16	H2 Forecast	Q4		11 May	

☐ Skip files with errors   OK   Cancel

Figure 2.31 – Table preview during the Combine step

This built-in combine function will allow you to append the **Sales** sheets from three different Excel files. This topic will be widely explored in *Chapter 5, Combining Queries for Efficiency*.

How it works...

The idea of this recipe was to show you the potential of the folder connector because often users end up connecting multiple times to single files and then perform an append step. This takes time, and it is difficult to maintain when the number of files becomes bigger.

The folder connector allows you to refresh your data and perform all Power Query operations automatically. If you add a file in your folder and click on refresh, you will end up with a final table enriched with data coming from this last file.

Creating a query from a database

This recipe shows how to connect to a database and how tables and views are displayed while selecting which tables to display and work with in Power Query.

You have two generic options:

- **Select tables or views** as you would see them with a database viewer such as SQL Server Management Studio.

- **Retrieve tables by writing SQL statements** in a specific section that will pop up.

Getting ready

In this recipe, in order to connect to a SQL database, you need to have an **Azure SQL Database instance** with **AdventureWorks** data, database credentials, or access through Azure Active Directory authentication (log in with your Microsoft account).

How to do it...

Once you open the Power BI Desktop application, you are ready to perform the following steps:

1. Go to **Get data**, click on **More**, and browse for **Azure SQL database**:

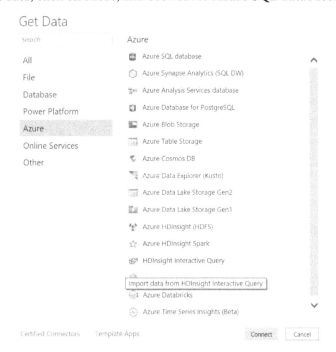

Figure 2.32 - Azure Connectors section

2. Enter the following information (expanding **Advanced options**):

a) **Server**: Server name

b) **Database**: `Adventureworks`

c) **Data Connectivity mode**: `Import`

d) **SQL statement**: This is a SQL view executed using the data source based on two tables in the database. The view is built as a SQL join between `FactResellerSales` and `DimSalesTerritory`:

```
SELECT s.[ProductKey]
      ,s.[SalesTerritoryKey]
      ,s.[SalesOrderNumber]
      ,s.[SalesOrderLineNumber]
      ,s.[RevisionNumber]
      ,s.[OrderQuantity]
      ,s.[UnitPrice]
      ,s.[ExtendedAmount]
      ,s.[UnitPriceDiscountPct]
      ,s.[DiscountAmount]
      ,s.[ProductStandardCost]
      ,s.[TotalProductCost]
      ,s.[SalesAmount]
      ,s.[OrderDate]
      ,p.[SalesTerritoryRegion]
      ,p.[SalesTerritoryCountry]
      ,p.[SalesTerritoryGroup]
  FROM [dbo].[FactResellerSalesXL_CCI] s
LEFT OUTER JOIN [dbo].[DimSalesTerritory] p ON
s.[SalesTerritoryKey] = [p.SalesTerritoryKey]
```

3. Copy and paste the code in the SQL statement section in order to get this view as the output table you will work on in Power Query:

Figure 2.33 – SQL Server database

4. Enter authentication details:

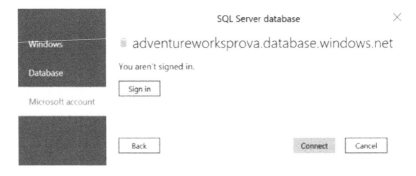

Figure 2.34 – SQL Server database authentication

5. After signing in, the output of the SQL statement will pop up as follows:

adventureworksprova.database.windows.net: Adventureworks

ProductKey	SalesTerritoryKey	SalesOrderNumber	SalesOrderLineNumber	RevisionNumber	OrderQuantity	UnitPrice	ExtendedAmou
469	2	SO788497	6	1	5	22.794	1
552	2	SO789266	6	1	5	54.894	2
558	2	SO789529	6	1	5	242.994	12.
460	2	SO789581	6	1	5	53.994	2.
555	2	SO789654	6	1	5	63.9	
272	2	SO789924	6	1	5	183.9382	91.
487	2	SO790341	6	1	5	32.994	1.
275	2	SO790673	6	1	5	356.898	17.
373	2	SO790711	6	1	5	1308.9375	6544.
353	2	SO790748	6	1	5	1391.994	69.
595	2	SO790846	6	1	5	338.994	16.
301	2	SO790869	6	1	5	714.7043	3573.
466	2	SO790914	6	1	5	14.1289	70.
374	2	SO791782	6	1	5	1466.01	73.
454	2	SO792033	6	1	5	35.994	1
490	2	SO793107	6	1	5	32.394	1.
553	2	SO793214	6	1	5	27.654	1.
299	2	SO793274	6	1	5	809.76	4.
227	2	SO793551	6	1	5	28.8404	14.
228	2	SO793618	6	1	5	29.994	1.

ⓘ The data in the preview has been truncated due to size limits.

‹   ›

OK   Cancel

Figure 2.35 – Table preview

6. Click on **Transform Data** in order to open the Power Query interface:

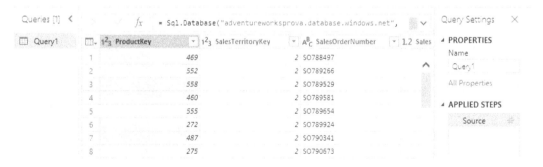

Figure 2.36 – Data preview in Power Query

7. Click on **Get data** and select the connector **Azure SQL Database**. In this case, we won't enter a SQL statement, but we will select an existing table in the database:

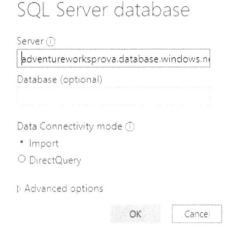

Figure 2.37 – SQL Server database connector

8. After signing in, a preview interface will appear, and you will be able to select the tables that you want to open in Power Query after clicking on **OK**:

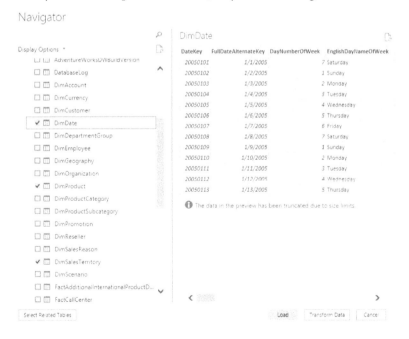

Figure 2.38 – SQL Database Navigator

9. You will see on the left a set of queries as an output of connecting directly to the database tables and writing a SQL statement querying the database as you would do with any other database viewing tool:

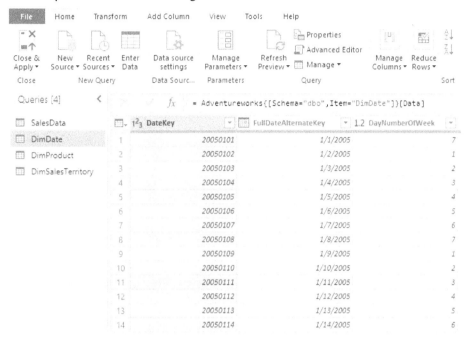

Figure 2.39 – Data preview in Power Query

If you open **Advanced Editor** for both types, you will notice that if you need to change the SQL code or you have to change the table name, you can do that directly from the **Advanced Editor** window:

a) The **SalesData** table's **Advanced Editor** code where you can see the details of the query run against the data source:

```
SalesData                                                        Display Options ▾    ?

    let
        Source = Sql.Database("adventureworksprova.database.windows.net", "Adventureworks",
            [Query="SELECT s.[ProductKey]#(lf),s.[SalesTerritoryKey]#(lf),s.[SalesOrderNumber]#(lf),s.
            [SalesOrderLineNumber]#(lf),s.[RevisionNumber]#(lf),s.[OrderQuantity]#(lf),s.[UnitPrice]#
            (lf),s.[ExtendedAmount]#(lf),s.[UnitPriceDiscountPct]#(lf),s.[DiscountAmount]#(lf),s.
            [ProductStandardCost]#(lf),s.[TotalProductCost]#(lf),s.[SalesAmount]#(lf)#(lf),s.
            [OrderDate]#(lf),p.[SalesTerritoryRegion]#(lf),p.[SalesTerritoryCountry]#(lf),p.
            [SalesTerritoryGroup]#(lf)FROM [dbo].[FactResellerSalesXL_CCI] s#(lf)LEFT OUTER JOIN [dbo]
            .[DimSalesTerritory] p ON#(lf)s.[SalesTerritoryKey] = p.SalesTerritoryKey"]}
    in
        Source

    ✓ No syntax errors have been detected.                        Done      Cancel
```

Figure 2.40 – Advanced Editor code for a SalesData query

b) The **DimDate** table's **Advanced Editor** code where you can see the details of the connection `Source`, the database retrieved from the server, `Adventureworks`, and the table selected from the data source `DimDate`:

DimDate

```
let
    Source = Sql.Databases("adventureworksprova.database.windows.net"),
    Adventureworks = Source{[Name="Adventureworks"]}[Data],
    dbo_DimDate = Adventureworks{[Schema="dbo",Item="DimDate"]}[Data]
in
    dbo_DimDate
```

Figure 2.41 – Advanced Editor code for the DimDate Query

How it works...

The Azure SQL Database connector also reflects how other database connectors work. If you connect to Amazon Redshift or an Oracle database, the experience will be very similar. Power Query provides a wide range of options for relational data sources and some of them may need the installation of specific drivers. For example, if you connect to SAP or Oracle, you have to install additional components (for example, in Oracle, the additional components will be the **Oracle Data Access Components** (**ODAC**)).

Creating a query from a website

Data is not only located in databases, but also in files, online services, and third-party applications as a growing number of users require the ability to connect to information available on the web. The idea behind the web connector is to allow easy and intuitive information extraction from websites. In this section, we will explore the possibilities of this connector and we will connect to a web page to extract data in an easily readable format.

Getting ready

For this recipe, you need Power BI Desktop and access to the following website: `https://www.packtpub.com/eu/all-products`.

How to do it...

In this recipe, the idea is to retrieve data from the Packt online catalog. By clicking on the preceding link, you will see the following site:

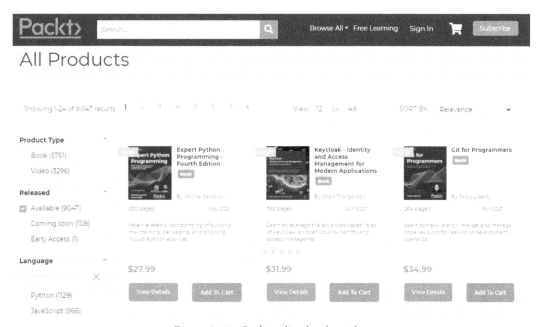

Figure 2.42 – Packt online book catalog

Imagine you want to extract data regarding the books available on this site.

Open Power BI Desktop and follow these steps:

1. Go to **Get data** and click on **Web**. Insert the link in the **URL** field:

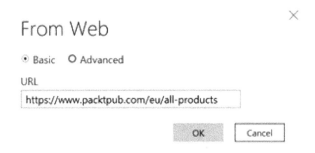

Figure 2.43 – Web connector

2. Authenticate as **Anonymous** (since it is a public website) and click on **Connect**:

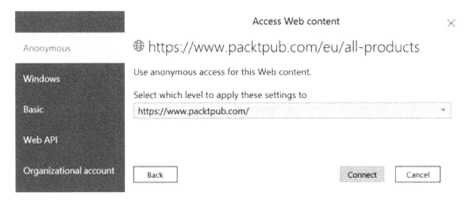

Figure 2.44 – Web connector authentication

3. After authenticating, the following preview window will pop up where, on the left, you can find a list of suggested tables and, on the right, you can see a data preview:

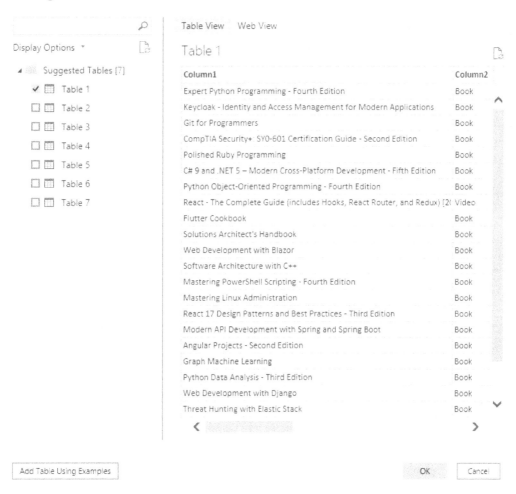

Figure 2.45 – Web tables preview

4. If you click on **Transform Data**, you will open the Power Query interface and then you can rename and clean up your data:

Figure 2.46 – Web data preview in Power Query

We will try another feature to extract data from the website and test an advanced link by inserting filters at the URL level:

1. Go to **Get data** and select the **Web** connector. Click on **Advanced** and split the URL `https://www.packtpub.com/eu/all-products?released=Available&tool=Azure&vendor=Microsoft` into three parts as in the next screenshot and click on **OK**:

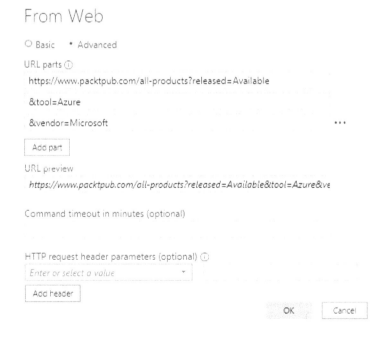

Figure 2.47 – Web connector advanced

2. The preview window will pop up. Click on **Add Table Using Examples**:

Add Table Using Examples

Figure 2.48 – Add Table Using Examples button

3. Start naming the columns as follows:

    a) **Title**

    b) **Author**

    c) **Nr. Pages**

    d) **Publication Date**

The columns should look like the ones in the following screenshot:

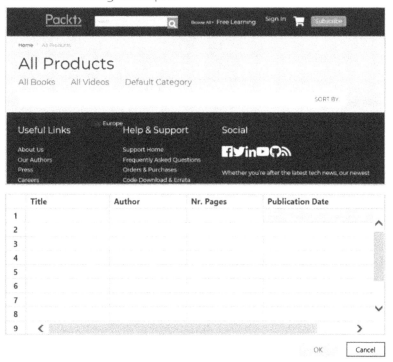

Figure 2.49 – Add Table Using Examples

4. Start filling in the first rows of each column and you'll see the other rows populate automatically:

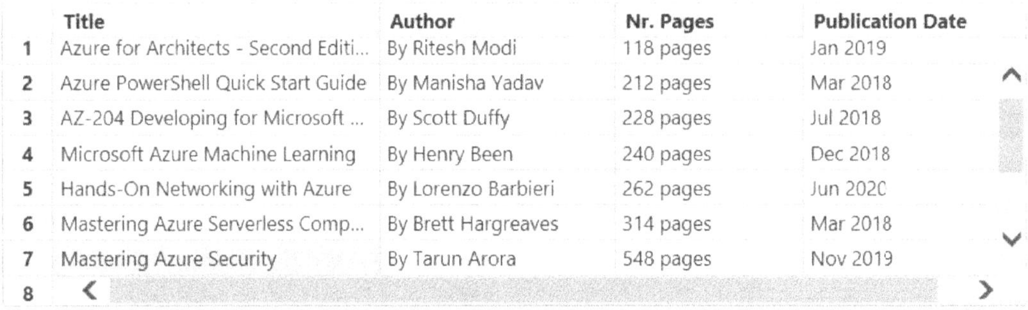
Title
Author
Nr. Pages
Publication Date
1 Azure for Architects - Second Editi... By Ritesh Modi 118 pages Jan 2019
2 Azure PowerShell Quick Start Guide By Manisha Yadav 212 pages Mar 2018
3 AZ-204 Developing for Microsoft ... By Scott Duffy 228 pages Jul 2018
4 Microsoft Azure Machine Learning By Henry Been 240 pages Dec 2018
5 Hands-On Networking with Azure By Lorenzo Barbieri 262 pages Jun 2020
6 Mastering Azure Serverless Comp... By Brett Hargreaves 314 pages Mar 2018
7 Mastering Azure Security By Tarun Arora 548 pages Nov 2019
8

Figure 2.50 – Add Table Using Examples details

5. Click on **OK** and you will generate a table within the **Custom Tables** section that you can select and load into Power Query:

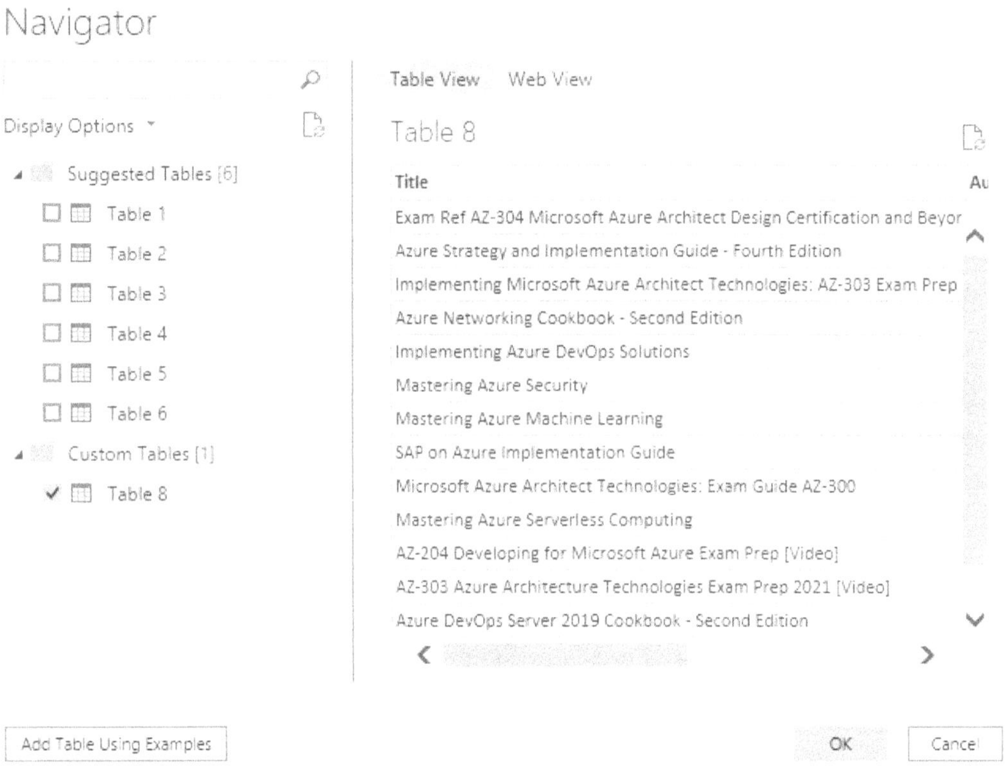
Navigator
Display Options
Suggested Tables [6]
Table 1
Table 2
Table 3
Table 4
Table 5
Table 6
Custom Tables [1]
Table 8
Table View
Web View
Table 8
Title
Au
Exam Ref AZ-304 Microsoft Azure Architect Design Certification and Beyor
Azure Strategy and Implementation Guide - Fourth Edition
Implementing Microsoft Azure Architect Technologies: AZ-303 Exam Prep
Azure Networking Cookbook - Second Edition
Implementing Azure DevOps Solutions
Mastering Azure Security
Mastering Azure Machine Learning
SAP on Azure Implementation Guide
Microsoft Azure Architect Technologies: Exam Guide AZ-300
Mastering Azure Serverless Computing
AZ-204 Developing for Microsoft Azure Exam Prep [Video]
AZ-303 Azure Architecture Technologies Exam Prep 2021 [Video]
Azure DevOps Server 2019 Cookbook - Second Edition
Add Table Using Examples
OK
Cancel

Figure 2.51 – Insert custom table from examples

With these simple steps, it is possible to connect and extract information from a website with a no-code approach. Users can focus on the content of data and not on the process of how to connect since Power Query allows them to do it in a few steps.

How it works...

This web connector not only allows users to connect to data from web pages by leveraging pre-defined tables identified by Power Query, but it also gives the ability to provide data examples from a web page and generate a custom table with relevant information for the user.

3
Data Exploration in Power Query

In a business context where data is acquiring more value, users—both business analysts and technical users—need a tool to easily explore their data. In order to get relevant insights from data that is continuously growing in amount and in terms of complexity, you need to have at your disposal a range of features and capabilities that will allow you to investigate aspects such as data size, quality, distribution, types, and other factors that will be covered in this chapter.

This chapter is focused on the data exploration features of Power Query. You will learn how to choose a subset of data and explore data profiling tools and query dependencies in order to see at a glance the data you will be dealing with. You will see how to smartly use Queries and Steps panes, with shortcuts and examples. Moreover, **Schema view** and **Diagram view** will be explained.

The recipes that will be explored in this chapter are listed here:

- Exploring Power Query Editor
- Managing columns
- Using data profiling tools
- Using Queries pane shortcuts
- Using Query Settings pane shortcuts
- Using Schema view and Diagram view

Technical requirements

For this chapter, you will require the following:

- **Power BI Desktop** (`https://www.microsoft.com/en-us/download/details.aspx?id=58494`)
- A **Power BI Pro** license and access to the `www.powerbi.com` portal
- A **Power BI workspace** on the **Power BI service**

The minimum requirements for installation are listed here:

- .NET Framework 4.6 (Gateway release August 2019 and earlier)
- .NET Framework 4.7.2 (Gateway release September 2019 and later)
- A 64-bit version of Windows 8 or a 64-bit version of Windows Server 2012 R2 with current **Transport Layer Security** (**TLS**) 1.2 and cipher suites
- 4 **gigabytes** (**GB**) disk space for performance monitoring logs

You can find the data resources referred to in this chapter at `https://github.com/PacktPublishing/Power-Query-Cookbook/tree/main/Chapter03`.

Exploring Power Query Editor

The aim of this recipe is to illustrate and describe the different sections of Power Query. The tool's design is studied to offer you an intuitive experience, and it is important to understand its logic to get the biggest benefit from it.

Getting ready

For this recipe, you need to have Power BI Desktop running on your machine.

How to do it...

Once you have opened Power BI Desktop, perform the following steps:

1. Click on **Transform data** to open an empty Power Query interface:

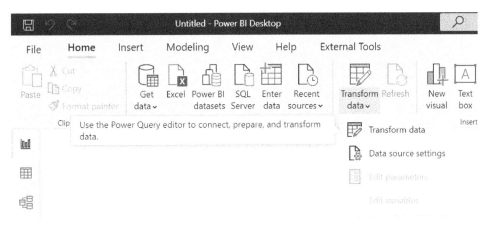

Figure 3.1 – Opening Power Query

2. Click on **Enter data** and create a table by creating two columns, ID and Value, and by entering values, as shown in the following screenshot:

Figure 3.2 – Entering data

3.   Click on **OK**.

The idea is to have an example query to illustrate the different Power Query **user interface** (**UI**) sections because, in this way, you will have all the different buttons under each tab activated.

It is possible to explore the Power Query interface to discover shortcuts and sections that will help you transform and enrich your data:

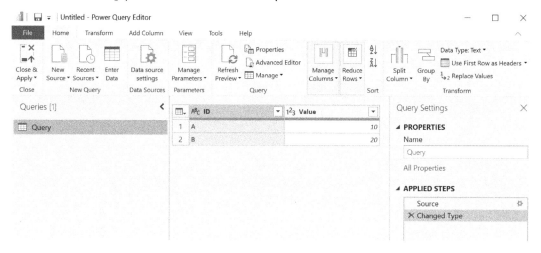

Figure 3.3 – Power Query UI

Generally, the UI can be divided into the following sections:

- The query ribbon includes six tabs:

   a) **Home**: This section consists of the most common tasks that you typically perform with Power Query. You can connect to a data source, expand data source settings, create and manage parameters, refresh queries' previews, and access the **Advanced Editor**.

   You can also perform some simple tasks such as selecting a subset of data (choosing columns and rows to be kept), splitting columns, replacing values, and grouping data.

   You can append and merge queries and access more advanced features, such as **AI Insights**.

b) **Transform**: This section includes tasks that transform existing data. You can pivot/unpivot columns and rows, replace values, apply calculations on columns, define data type formats, and you can also run Python and R scripts on data in the selected query:

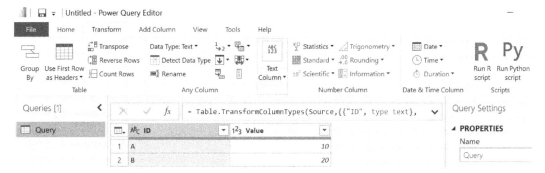

Figure 3.4 – Transform tab

c) **Add Column**: This tab enables you to enrich data and add new information that originates from existing data. It is possible to add conditional or custom columns, extract values, and apply numeric calculations. From this section, you can apply **AI Insights**:

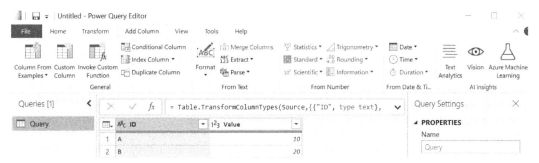

Figure 3.5 – Add Column tab

d) **View**: This section allows you to enable or disable the view of certain sections of the Power Query UI—for example, the **Query Settings** pane and the **Formula Bar**. Tools for data quality analysis are located under this tab, and the **Advanced Editor** can be accessed from here as well. Moreover, it is possible to open the **Query Dependencies** window:

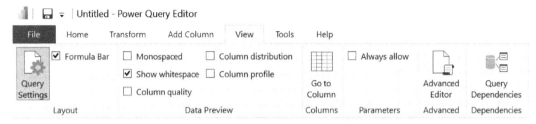

Figure 3.6 – View tab

e) **Tools**: This section contains **Diagnostics** tools:

Figure 3.7 – Tools tab

f) **Help**: This tab allows you to have quick access to learning resources, Power Query documentation, and community blogs and web pages:

Figure 3.8 – Help tab

Additionally, in the **Queries** section in the left-hand pane, you can find a list of the current Power Query session queries.

The central table view shows the data from the selected query. You can perform some steps directly from the table view by right-clicking on the column name:

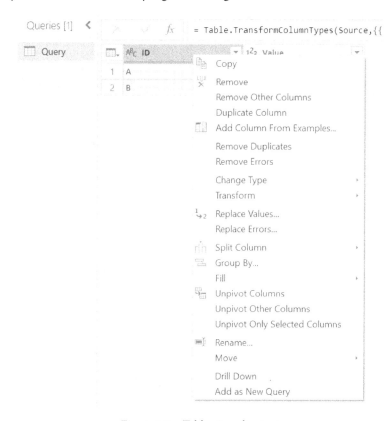

Figure 3.9 – Table view shortcut

The **Query Settings** section on the right shows all the applied steps that you do while transforming and enriching data. In this way, every data type change, new column addition, or new calculation performed can be traced.

> **Note**
> It is important that you are aware of how the UI is organized in order to be able to quickly find and access the transformations you need.

Managing columns

After connecting to a data source from Power Query and after having selected the table to which it is connected, it is best practice to reduce and delete all data that is not relevant for the preparation and transformation processes and therefore for reporting.

You have the possibility to choose the columns you want to work with, thereby reducing the amount of data involved. With this recipe, we will see how to quickly and intuitively select columns in order to speed up the data preparation process.

Getting ready

For this recipe, you need to download the `FactResellerSales` **comma-separated values (CSV)** file into your local folder.

In this example, we will refer to the `C:\Data` folder.

How to do it...

Once you have opened your Power BI Desktop application, you are ready to perform the following steps:

1. Click on **Get Data** and select the **Text/CSV** connector.

2. Browse to your local folder where you downloaded the `FactResellerSales` CSV file and open it. A window with a preview of the data will pop up; click on **Transform data**.

3. Within the **Home** tab, focus on the **Manage Columns** section:

Figure 3.10 – Manage Columns

You have two possibilities to restrict the number of columns:

- **Choose Columns**—Click this to choose columns you wish to keep

- **Remove Columns**—Click this to remove columns you do not need

Choosing columns

You can choose columns you wish to keep with the following steps:

1.   Click on **Choose Columns**:

Figure 3.11 – Choose Columns button

2.   A view where you can choose the columns you want to keep will appear:

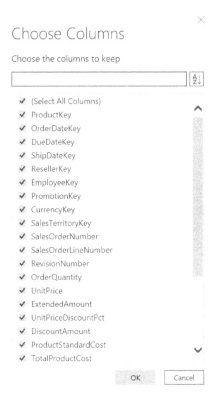

Figure 3.12 – Choose Columns window

3. Remove the flag from **(Select All Columns)**, type `Sales` in the search bar, and flag **(Select All Search Results)**:

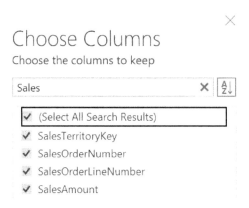

Figure 3.13 – Choose Columns selection

4. Repeat the same step by selecting all columns containing `Date` in their name and click on **OK**; you will end up having 10 columns instead of 27.

The **Choose Columns** section is also aimed at browsing tables more quickly and finding the column you want to transform or enrich. If you click on **Go to Column**, a relative window pops up whereby you can change which column you end up selecting:

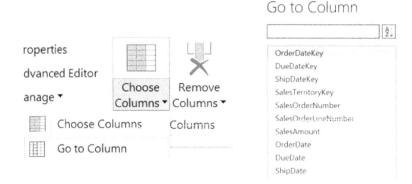

Figure 3.14 – Go to Column

While the **Choose Columns** step allows you to add a Power Query step and transforms your query, the **Go to Column** step is just a UI function for easy navigation and does not result in a query step.

Removing columns

You can also decide to do this the other way around and not choose columns to keep, but rather delete columns by completing the following steps:

1. Select the first three columns by pressing the *Ctrl* button and clicking on each in order to get the following view:

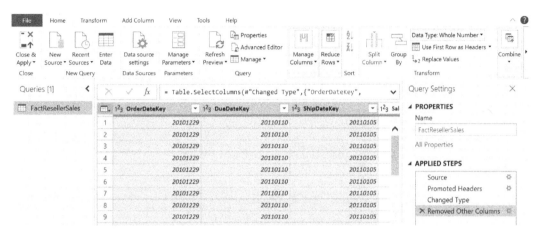

Figure 3.15 – Column selection

2. Expand the **Remove Columns** button and click on **Remove Columns** to remove the selected columns:

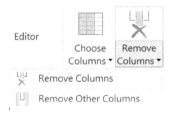

Figure 3.16 – Remove Columns button

If you had clicked on **Remove Other Columns**, you would have kept the selected columns and instead removed the others.

These flexible actions are useful because they allow you to choose how to remove unnecessary data and optimize the data preparation process. Useless and redundant data tends to slow down the entire data transformation pipeline, and it is better to discard everything that is not useful and strategic for data analysis.

Remember that you can always go back and modify the **Remove Columns** step in the **Apply Steps** pane. In fact, **Choose Columns** decides by itself whether selecting or removing (other) columns is the most efficient action.

Using data profiling tools

You may deal with great amounts of data and need tools that allow you to quickly check data quality and distribution and get insights from columns' profiles.

Power Query offers an intuitive way of exploring data to identify *bad* data. Data profiling is especially convenient when you are working with large volumes of data and you want to quickly visualize the composition of that data.

Getting ready

For this recipe, you need to download the `FactInternetSales2` CSV file into your local folder.

In this example, we will refer to the `C:\Data` folder.

How to do it...

Once you have opened your Power BI Desktop application, you are ready to perform the following steps:

1. Click on **Get Data** and select the **Text/CSV** connector.

2. Browse to your local folder where you downloaded the `FactInternetSales2` CSV file and open it. A window with a preview of the data will pop up; click on **Transform data**.

3. Browse the query ribbon and click on **View**:

Figure 3.17 – View tab

4. Flag **Column quality** and observe the results:

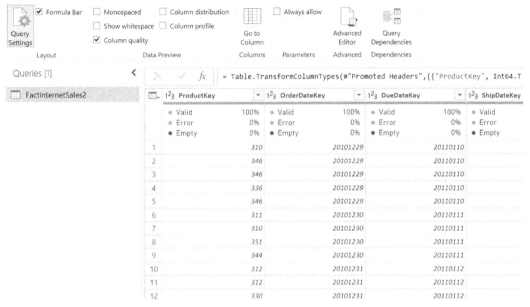

Figure 3.18 – Column quality

This view shows three categorizations that define the content of the column in terms of *quality*, expressed in percentages:

- **Valid**: The percentage of *valid* data according to column data type

- **Error**: The percentage of rows with errors

- **Empty**: The percentage of empty rows

This information is based on the top 1000 rows, as you can observe at the bottom of the Power Query UI:

26 COLUMNS, 999+ ROWS   Column profiling based on top 1000 rows

Figure 3.19 – Profiling based on the top 1000 rows

5. Click on **Column profiling based on top 1000 rows** and select **Column profiling based on entire data set**:

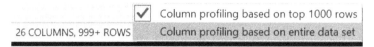

Figure 3.20 – Profiling based on entire dataset

6. Now, observe the new values under the column names:

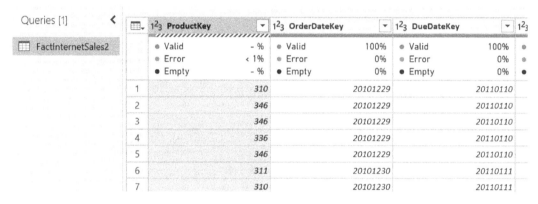

Figure 3.21 – Column quality on entire dataset

You will see different values from the ones based on the preview data. It is always important to check and not rely exclusively on the data quality profiled on the top 1000 rows.

7. Next, using your cursor, hover under the ProductKey column and see that a tooltip will appear:

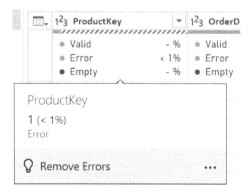

Figure 3.22 – Tooltip on column quality

You can see that there is a value that is compromising the column quality. In order to manage this error, we have to perform a set of actions. You could directly click on **Remove Errors** and remove the row affected by that error, but in this way, you could lose relevant information.

8. Click on the three dots (**…**) and click on **Keep Errors**:

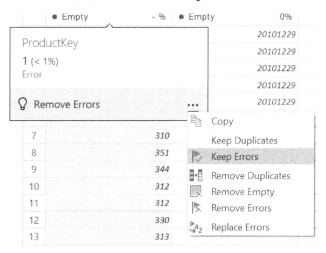

Figure 3.23 – Keep Errors

9. In this way, you will be able to filter to see the rows affected by errors:

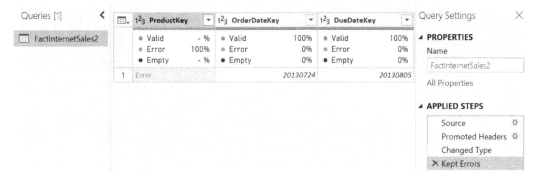

Figure 3.24 – Rows affected by errors

10. Click on the value contained in the `ProductKey` column, and you will see the value that is causing the error:

Figure 3.25 – Error in data quality

You will see that the error was caused by a numeric value contaminated by a letter, and this error comes from the data source. It is possible to fix this at the Power Query level without changing the value on the data source, but by having a correct value for reporting purposes.

11. In order to fix this value, you can go back to the previous view with all data by deleting the **Kept Errors** and **ProductKey** steps from the **APPLIED STEPS** pane:

Figure 3.26 – Deleting applied steps

12. Select the `ProductKey` column, right-click on it, and click on **Replace Errors…**:

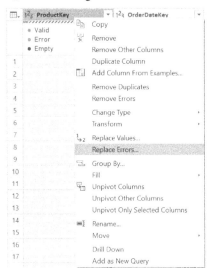

Figure 3.27 – Replace Errors

13. Enter the value 480 in order to replace the error that we know to be 480b, as retrieved from previous steps:

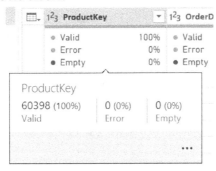

Figure 3.28 – Replace Errors window

14. After this replacement is applied, you can see that now, from a data quality point of view, the column has no errors:

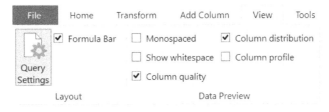

Figure 3.29 – Column quality

Next, we will focus on the **Column distribution** feature that provides additional information to the **Column quality** feature.

In order to see how this tool works, follow the next steps:

1. Browse the query ribbon, click on **View**, and flag **Column distribution**:

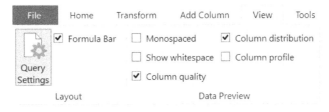

Figure 3.30 – Column distribution flagged

2. You will see a section that shows a number of distinct and unique values in a column and a visualization showing the distribution of these values:

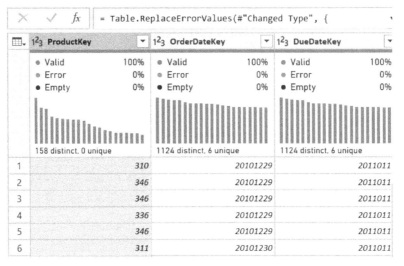

Figure 3.31 – Column distribution section

3. If you hover with your cursor on the section we are considering, you can see details on how many distinct and unique values that column has:

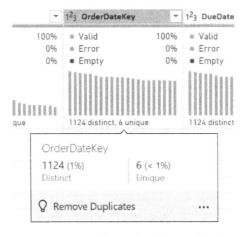

Figure 3.32 – Column distribution details

Finally, we will focus on the **Column profile** feature. In order to see how to leverage this tool, follow the next steps:

1. Browse the query ribbon, click on **View**, and flag **Column profile**:

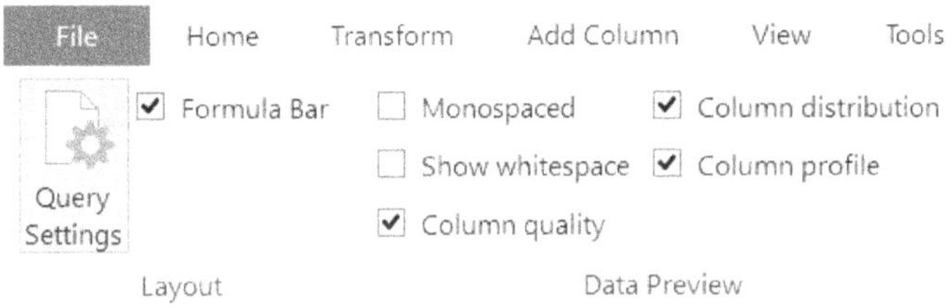

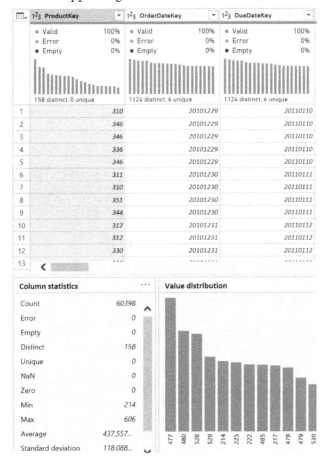

Figure 3.33 – Column profile flagged

2. You will see a section appearing at the bottom of the UI, as follows:

Figure 3.34 – Column profile section

This section gives additional details on the columns' content. On the left, you can see data other than that seen under **Column quality** and **Column distribution**, such as **NaN** values, **Min** and **Max**, **Average**, **Standard deviation**, and a count of even and odd numbers.

On the right, you can see a column chart with detailed values on data distribution.

3. If you click on the three dots (**…**), you will see that you can choose different grouping types according to the column's data type:

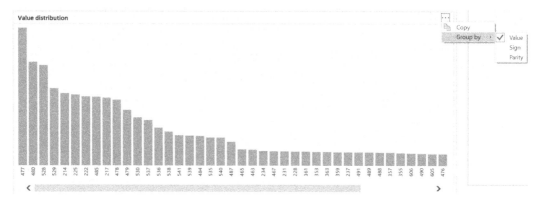

Figure 3.35 – Column distribution chart

4. If you hover with your cursor on the chart, a tooltip will appear with additional information on that column's values; if you click on the three dots (**…**), you can apply directly from here a set of transformations such as filtering data or replacing values:

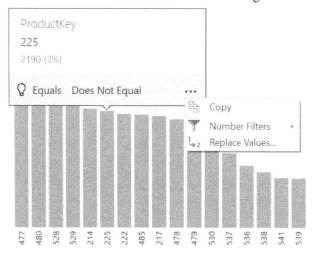

Figure 3.36 – Column distribution tooltip details

With this set of tools, a user can easily get relevant insights from queries with only a few clicks. This is useful when you have to make corrections once you have identified errors that cannot be quickly corrected at a data source level.

Using Queries pane shortcuts

The Power Query UI offers smooth data navigation. This recipe will help you to understand how to use the **Queries** pane on the left of the UI. Other than basic query navigation, it is possible to perform different actions by using some shortcuts.

Getting ready

For this recipe, you need to download the FactInternetSales CSV file into your local folder.

In this example, we will refer to the C:\Data folder.

How to do it...

Once you open your Power BI Desktop application, you are ready to perform the following steps:

1. Click on **Get Data** and select the **Text/CSV** connector.

2. Browse to your local folder where you downloaded the FactInternetSales CSV file and open it. A window with a preview of the data will pop up; click on **Transform data**.

3. Go to the **Queries** pane and right-click on the query:

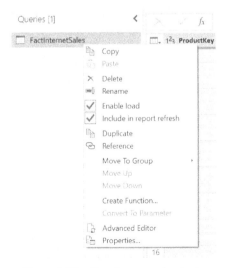

Figure 3.37 – Queries pane shortcuts

4. From this view, you can carry out the following actions:

a) **Copy** and **Paste** queries.

b) **Delete** queries.

c) **Rename** queries.

d) **Enable load**: This allows you to enable or disable a query to be loaded in the model/dataflow.

e) **Include in report refresh**: This allows you to include/exclude a query in a model refresh.

f) **Duplicate** queries.

g) **Reference**: This enables you to create a new query that uses the applied steps from the query it is referencing. It does not duplicate a query, and every change in the original query will be reflected in the one that is referencing it.

h) **Move To Group**: You can create groups and organize your queries in folders.

i) **Move Up/Move Down** queries.

j) **Create Function:** This allows to create functions on top of the selected query.

k) **Convert to Parameter**: This allows you to convert queries into parameters.

l) **Advanced Editor**: From this shortcut, you can access the **Advanced Editor** view of the selected query.

m) **Properties…**: You can access a **Properties** section where you can rename a query, add a description, and flag the properties you see in the following screenshot:

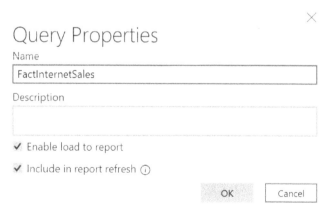

Figure 3.38 – Queries properties

Through the **Queries** pane, you can quickly perform some common tasks and organize queries, especially when their number and complexity grow.

Using Query Settings pane shortcuts

One of the main benefits of Power Query is that every transformation is mapped and traced in the **APPLIED STEPS** pane. This is part of the **Query Settings** section, where you can see all the transformations and can perform some actions through a set of shortcuts.

The aim of this recipe is to explore the **Query Settings** section and explore possible activities.

Getting ready

For this recipe, you need to download the `FactInternetSales` CSV file into your local folder.

In this example, we will refer to the `C:\Data` folder.

How to do it...

Once you open your Power BI Desktop application, you are ready to perform the following steps:

1. Click on **Get Data** and select the **Text/CSV** connector.

2. Browse to your local folder where you downloaded the `FactInternetSales` CSV file and open it. A window with a preview of the data will pop up; click on **Transform data**.

3. Filter on the `ProductKey` column and select the values `310`, `311`, and `312`:

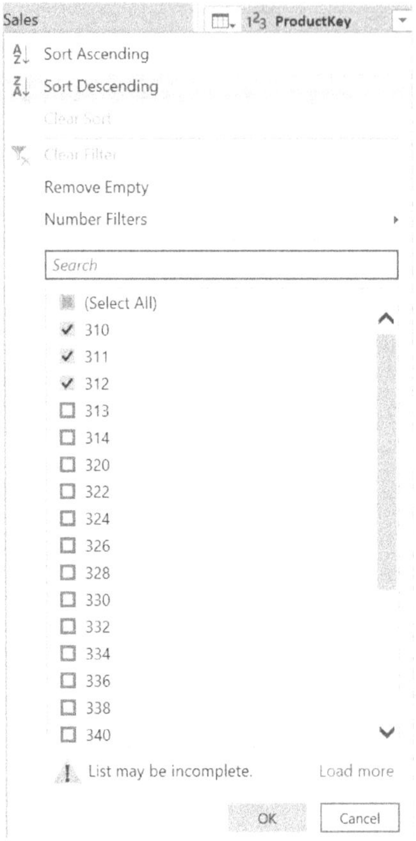

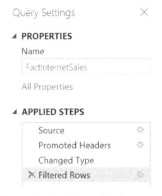

Figure 3.39 – Filtering on ProductKey

4. Go to the **Query Settings** pane and observe the **PROPERTIES** and **APPLIED STEPS** sub-sections:

Figure 3.40 – Query Settings pane

5. From the **PROPERTIES** sub-section, you can rename your query, and if you click on **All Properties**, the following window will pop up:

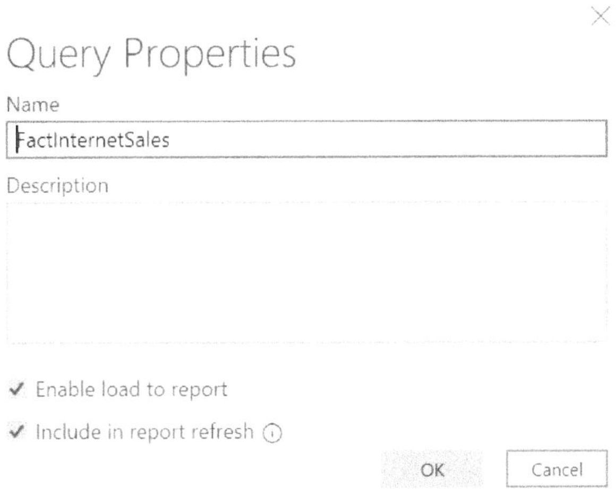

Query Properties

Name

FactInternetSales

Description

✓ Enable load to report

✓ Include in report refresh ⓘ

OK   Cancel

Figure 3.41 – Query Properties window

As already seen in the *Using Queries pane shortcuts* recipe, in this view you can rename a query, enter a query description, and flag/unflag some additional properties.

6. Focus now on the **APPLIED STEPS** sub-section, and right-click on the **Filtered Rows** step:

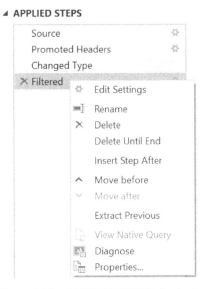

Figure 3.42 – APPLIED STEPS shortcuts

7. From this view, you can carry out the following actions:

a) **Edit Settings**: Make changes within a step.

b) **Rename** a step by giving it a more explicit or intelligible name.

c) **Delete** a step.

d) **Delete Until End**: Delete all steps until the end, including the one selected. An intermediate window will appear, asking if you are sure you want to perform this action.

e) **Insert Step After** the step you have selected.

f) **Move before/Move after** the step selected. You can achieve the same by dragging and dropping the step you want to move.

g) **Extract Previous** queries if you want to transfer a set of transformations to a new query. When you click on this, a window will appear, where you will be required to name the new query.

h) **View Native Query**: When clickable, you can see the query statements that are running against your data source.

i) **Diagnose**: You can click this to run a detailed analysis for that particular step and get diagnostics insights.

j) **Properties…**: A window will pop up where you can rename a step and add a description.

Using Schema view and Diagram view

You often need to visualize tables and columns focusing on a data schema and with the aim of performing transformations at a metadata level. Using the traditional **Power Query** view may end up being slower because the calculations are performed for all displayed rows, both the preview or the entire dataset. Another need is related to having a visual way to transform data whereby it is easier to understand the data preparation flow.

This recipe aims to show how to leverage the recently introduced **Schema view** and **Diagram view** available in Power Query Online.

Getting ready

For this recipe, you need to have access to the Power BI service and to have an existing workspace.

You need also to connect to an **Azure SQL database** with **AdventureWorks** data. You need to have access to a running database.

How to do it...

After you log in to the Power BI portal, perform the following steps:

1. Browse to your workspace, click on **New**, and click on **Dataflow**:

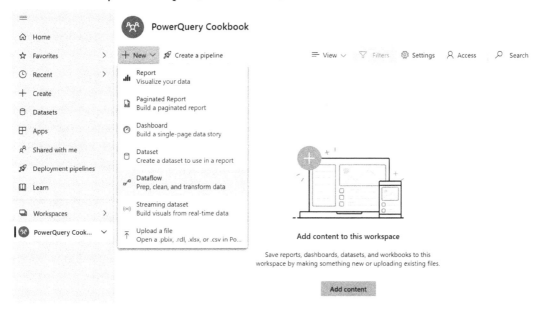

Figure 3.43 – Creating a dataflow

2. Click on **Add new tables**:

Figure 3.44 – Add new tables button

3.   Search for **Azure SQL database** and click on the connector:

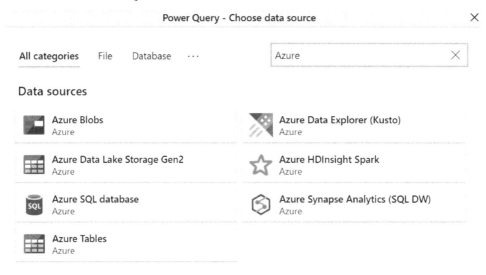

Figure 3.45 – Azure SQL database connector

4.   Enter the server and database name and authentication details:

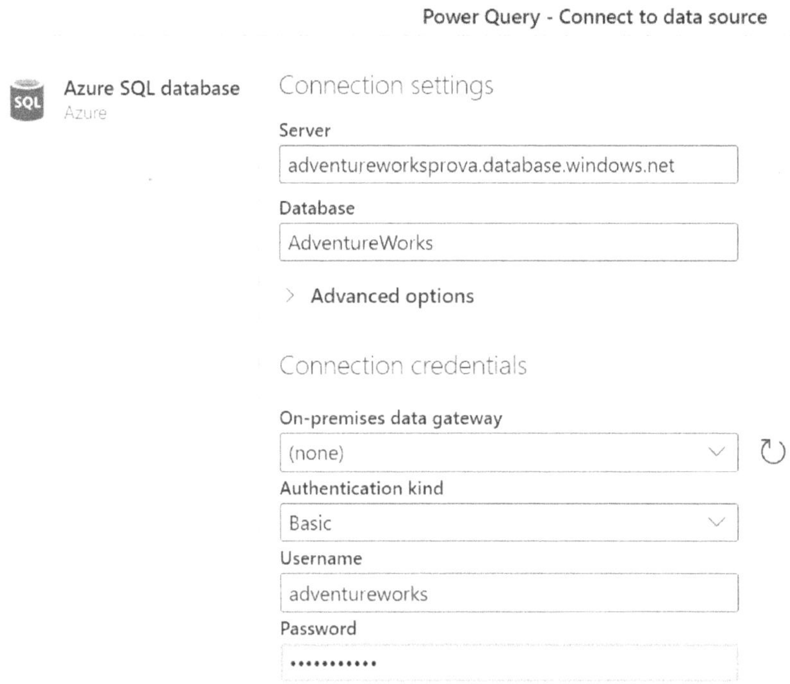

Figure 3.46 – Azure SQL database details

5. Select the `DimGeography` and `FactInternetSales` tables and click on **Transform data**:

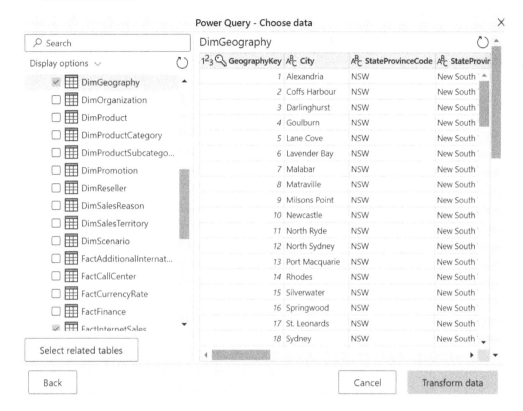

Figure 3.47 – Data preview

6. Browse to the **View** tab and click on **Schema view**:

Figure 3.48 – Schema view button

7.   Data information will be viewed in the following mode:

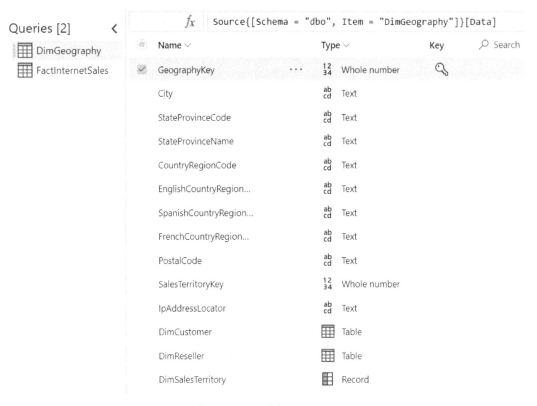

Figure 3.49 – Schema view

8. Activities that you can perform thanks to **Schema view** include the following:

a) Reordering columns with drag and drop:

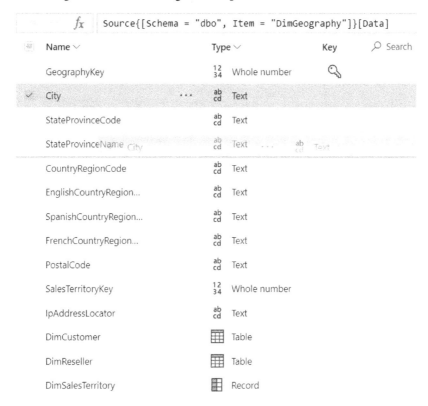

Figure 3.50 – Reordering a column in Schema view

b) Applying a transformation listed on the **Schema tools** tab:

Figure 3.51 – Schema tools tab

As you can see, you can perform varied tasks such as choosing or removing columns, changing data types, marking a column as a key, duplicating, and renaming. You can also use other transformations available in other tabs. You can always go back to **Data view** by closing **Schema view**.

In general, this view is useful when you want to speed up some transformation and focus only on metadata. Once you close **Schema view**, you will apply the steps together, which ends up being more efficient.

You can also use a visual method to apply transformations to your data and leverage a feature available in Power Query Online: **Diagram view**.

Let's see how you can use this feature by following the next steps:

1. Once you have closed **Schema view** from the previous example, let's open **Diagram view**. Browse to the **View** tab and click on **Diagram view**:

Figure 3.52 – Diagram view button

2. Once you have turned to **Diagram view**, you will see that each query is defined by a block, as illustrated in the following screenshot. You can also see a table preview:

Figure 3.53 – Diagram view queries

You can click on the **Expand** icon and visualize steps applied to each query:

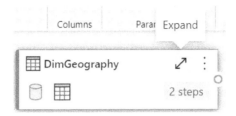

Figure 3.54 – Expand option in DimGeography

3. After you have expanded the query, you can click on the plus (+) icon and navigate and choose the transformation step you want to perform:

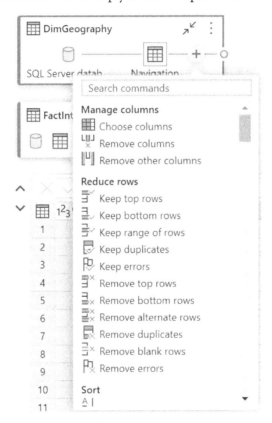

Figure 3.55 – Adding a new step in Diagram view

4.   Click on **Choose columns** (the first option you see), select columns as shown in the following screenshot, and click on **OK**:

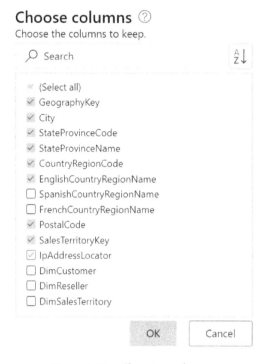

Figure 3.56 – Choosing columns

5.   Add another step by clicking on the plus (**+**) icon and select **Merge queries as new**:

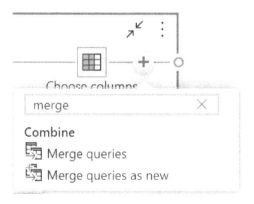

Figure 3.57 – Merge queries as new

6. Select the `SalesTerritoryKey` column from `DimGeography` as **Left table for merge** and from `FactInternetSales` as **Right table for merge**. Then, select **Left outer** in the **Join kind** field and click on **OK**:

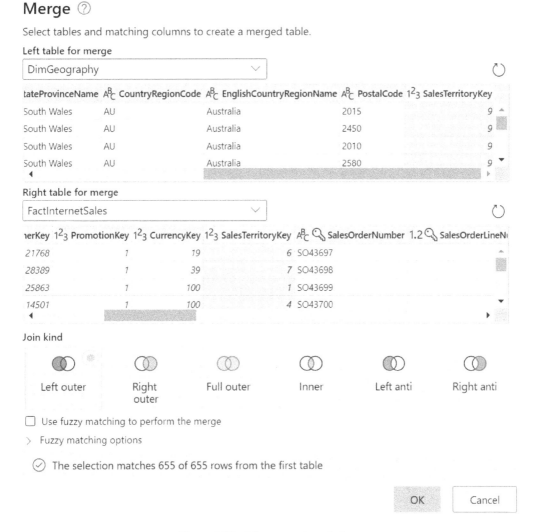

Figure 3.58 – Merge queries window

7. Click on **Highlight related queries** and see how queries involved with the **Merge** operation are highlighted:

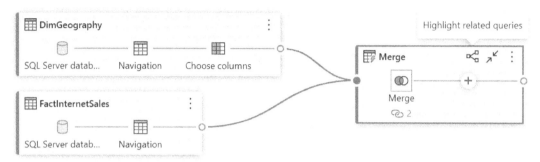

Figure 3.59 – Merging in Diagram view

In general, **Diagram view** is useful for understanding the flow of data, seeing quickly how queries, parameters, and lists are related to each other, and easily expanding details for every step.

4

Reshaping Your Data

When data grows in size and complexity, you need tools that allow you to make sense of your data and create views that can be relevant for reporting and presenting that data. In order to do this, they need features and options to reshape their data properly and to clean it if needed, especially when you can't edit data sources directly. Often, they connect to data that does not have a correct schema or a correct column name and they need to group and change how data is displayed.

In the recipes in this chapter, you will see how to leverage **Power Query**'s built-in features that will allow you to reshape your data, to change its structure, and to adapt it to your reporting needs.

In this chapter, we will cover the following recipes:

- Formatting data types
- Using first rows as headers
- Grouping data
- Unpivoting and pivoting columns
- Filling empty rows
- Splitting columns
- Extracting data

- Parsing JSON or XML

- Exploring artificial intelligence insights

Technical requirements

For this chapter, you will be using the following:

- Power BI Desktop: `https://www.microsoft.com/en-us/download/details.aspx?id=58494`

- Power BI Premium capacity or a Premium Per User license

The minimum requirements for installation are as follows:

- **.NET Framework 4.6** (Gateway release August 2019 and earlier)

- **.NET Framework 4.7.2** (Gateway release September 2019 and later)

- A 64-bit version of Windows 8 or a 64-bit version of Windows Server 2012 R2 with current TLS 1.2 and cipher suites

- 4 GB disk space for performance monitoring logs

You can find the data resources referred to in this chapter at `https://github.com/PacktPublishing/Power-Query-Cookbook/tree/main/Chapter04`.

Formatting data types

Data analysts and business intelligence users generally don't have editing permissions on data sources. You can't change data types directly on the database. Asking for custom changes may require time and it can be complex. In this sense, Power Query becomes a powerful tool because it helps you define and customize data types. In this recipe, we will see different options on how to change data types of specific columns or of entire tables.

Getting ready

For this recipe, you need to have **Power BI Desktop** running on your machine. You need to download the following files in a local folder:

- `FactResellerSales` CSV file

- `FactInternetSales` CSV file

In this example, we will refer to the `C:\Data` folder.

How to do it

Once you open your Power BI Desktop application, you are ready to perform the
following steps:

1. Click on **Get data** and select the **Text/CSV** connector:

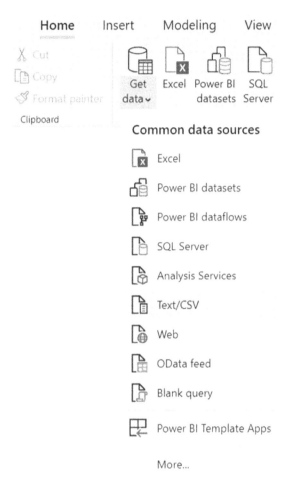

Figure 4.1 – Text/CSV connector

2. Browse to your local folder where you downloaded the `FactInternetSales`
   CSV file and open it. The following window with a preview of the data will pop up.
   Click on **Transform Data**:

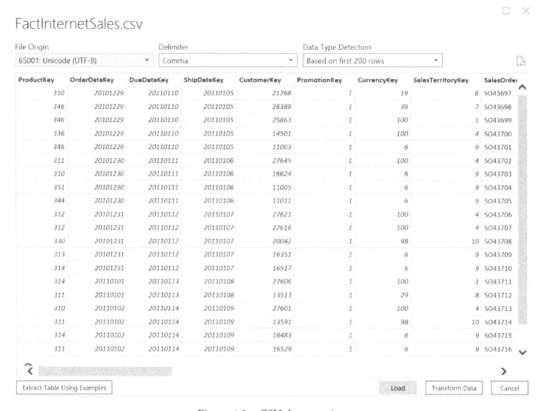

Figure 4.2 – CSV data preview

3. The usual Power Query interface will appear with data displayed. Focus on the
   APPLIED STEPS pane and see that two steps were applied automatically:

   a) **Promoted Headers:** The first row of the file is promoted as the columns' header.

   b) **Changed Type**: For unstructured data sources such as CSV/TXT, this step is
      applied by default where Power Query detects the most adequate data type:

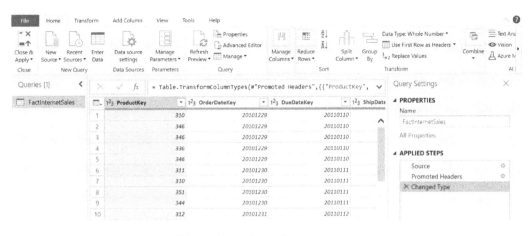

Figure 4.3 – Changed data type step

If you focus on the selected column, `ProductKey`, you can see that the data type is displayed on the left of the column name as shown in the following screenshot:

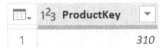

Figure 4.4 – Data type

4.   If you click on the data type icon, a wider selection of data types will expand:

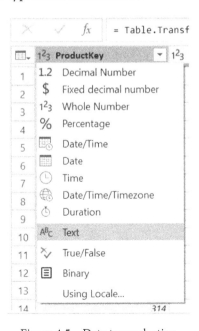

Figure 4.5 – Data type selection

Click on **Text** in order to convert this column type from **Whole Number** to **Text**.

5. The **Change Column Type** window will pop up because we added a change type to the one generated by default:

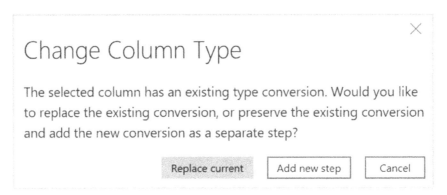

Figure 4.6 – Change Column Type

You have different options to change data types for one or more columns:

- **Replace current**: You can replace the current step with the one defined in the previous section. You will cancel the data type detection performed by Power Query when you loaded the data and apply the step you defined for that single column. You will then have to define other columns' data types.

- **Add new step**: You can add a new step and keep the one defined automatically.

In this case, we will select the last one in order to keep the data types detected by the **Changed Type** step. Click on **Add new step**. This happens because when you are loading date/time values from a text file and you want to convert to date, Power Query will not do that directly; you have to first change from text to date/time (which is done automatically by data type detection) and as a new step, convert to date.

Now let's see another example that shows how to manipulate and change data types as follows:

1. Click on **File**, **Options and settings**, and then on **Options**:

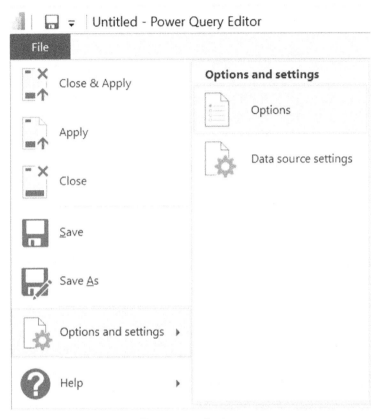

Figure 4.7 – Options

2. Focus on the **Data Load** tab and on the **Type Detection** section. You can disable or enable the automatic data type detection from this window. Flag the third option to disable the automatic detection as shown in the following screenshot and then click on **OK**:

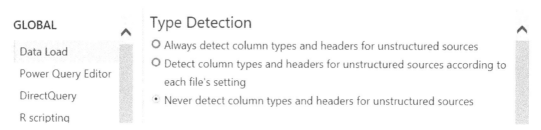

Figure 4.8 – Data type detection setting

3. Click on **Get Data** and select the **Text/CSV** connector.

4. Browse to your local folder where you downloaded the FactResellerSales CSV file, open it, and click on **OK**.

5. You will see that in the Power Query interface, you won't see automatic applied steps because we disabled automatic data type detection:

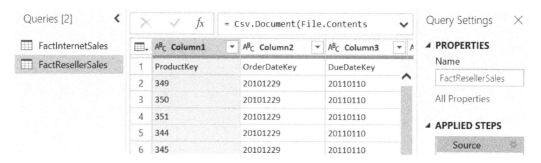

Figure 4.9 – Table preview

6. Browse to the **Transform** tab and click on **Use First Row as Headers**:

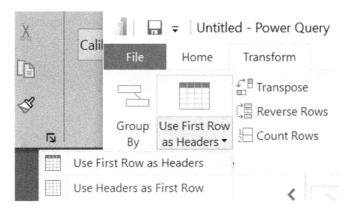

Figure 4.10 – Use First Row as Headers

7.   Then select all columns and click on **Detect Data Type**:

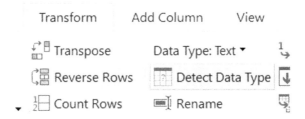

Figure 4.11 – Detect Data Type

8.   You will see that data types were detected:

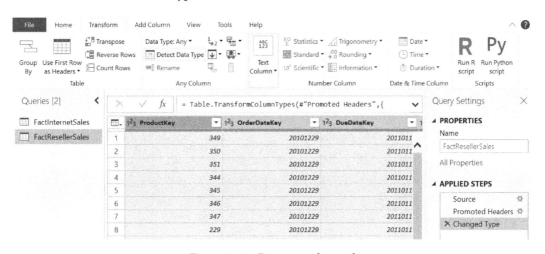

Figure 4.12 – Data types detected

9. You will see that some values were not detected correctly because number values were detected as text:

A^B_C ProductStandardCost	A^B_C TotalProductCost	A^B_C SalesAmount	A^B_C TaxAmt	A^B_C Freight
1898,0944	1898,0944	2024,994	161,9995	50,6249
1898,0944	5694,2832	6074,982	485,9986	151,8746
1898,0944	1898,0944	2024,994	161,9995	50,6249
1912,1544	1912,1544	2039,994	163,1995	50,9999
1912,1544	1912,1544	2039,994	163,1995	50,9999
1912,1544	3824,3088	4079,988	326,399	101,9997
1912,1544	1912,1544	2039,994	163,1995	50,9999

Figure 4.13 – Data types not detected correctly

10. Select the columns you see in the preceding screenshot: ProductStandardCost, TotalProductCost, SalesAmount, TaxAmt, and Freight. Right-click on one of the column names, click on **Change Type**, and then click on **Using Locale…** as you can see in the following screenshot:

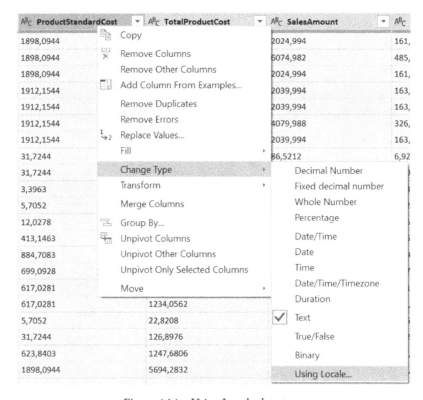

Figure 4.14 – Using Locale data type

11. A window will pop up where you can define local number formats. This feature is useful when you have to deal with the decimal separator, which has different formats according to the country settings set on the machine or at the tool level. Select **Fixed decimal number** as **Data Type** and **Locale** as **English (Germany)** since we want to use a comma as decimal separator, and click on **OK**:

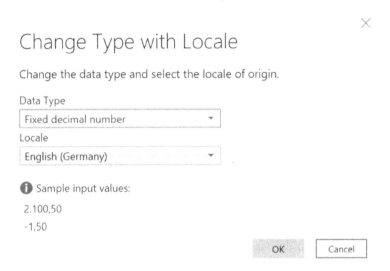

Figure 4.15 – Define data type locale

12. Observe how the selected columns now show the correct data types:

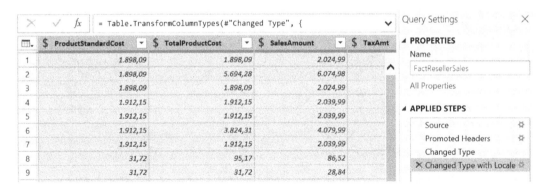

Figure 4.16 – Data types corrected

There are different options on how to define data types. You can rely on automatic detection; you can customize each column and deal with customized formats. This flexibility can be achieved at the table or at column level by selecting single or subsets of columns.

Using first rows as headers

When working with unstructured data, there is no data structure defined and the schema has to be determined at the Power Query level. In the previous recipe, we saw how data type is detected, while in this one we will see how you can define columns' headers.

Getting ready

In this recipe, you need to download the following file in a local folder:

- `FactResellerSales2` CSV file

In this example, we will refer to the `C:\Data` folder.

How to do it

Once you open your Power BI Desktop application, you are ready to perform the following steps:

1. Click on **Get Data** and select the **Text/CSV** connector.

2. Browse to your local folder where you downloaded the `FactResellerSales2` CSV file and open it. A window with a preview of the data will pop up; click on **Transform Data**.

3. The usual Power Query window will pop up and it is easy to note that this data needs to be cleaned because we do not have headers and the first row contains data not useful for any kind of analysis, as seen in the following screenshot:

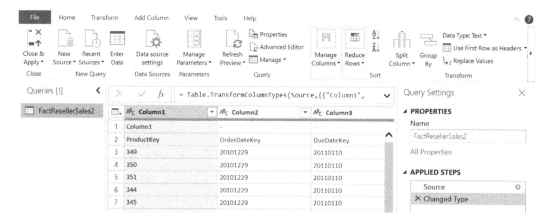

Figure 4.17 – Power Query data interface

4. Click on **Remove Rows** and then on **Remove Top Rows**. In the **Remove Top Rows** window, insert 1 to remove the row we don't need and click on **OK**:

Figure 4.18 – Remove Top Rows

5. Now click on **Use First Row as Headers**:

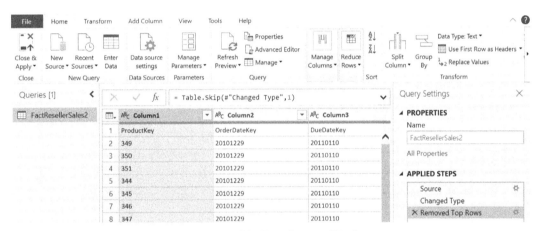

Figure 4.19 – Use First Rows as Headers

6. You can see that now we have a defined schema with correct column headers:

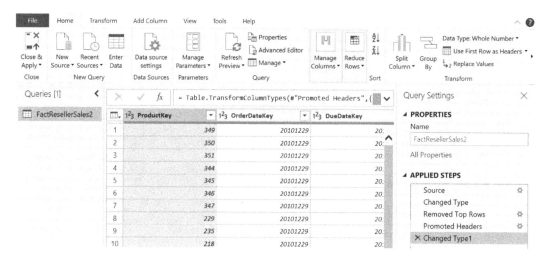

Figure 4.20 – Cleaned data

When promoting rows to column headers, you will see that a **Changed Type1** step is applied automatically.

Grouping data

We connect to a wide variety of data and usually connect to data with high levels of detail that it may not need for reporting. Instead of loading all data, we can define relevant aggregations and group data according to custom logic at the Power Query level. In this recipe, we will see how to define grouping logic and how to aggregate data easily.

Getting ready

For this recipe, you need to have Power BI Desktop running on your machine. You need to download the following file in a local folder:

- `FactInternetSales` CSV file

In this example, we will refer to the `C:\Data` folder.

How to do it

Once you open your Power BI Desktop application, you are ready to perform the following steps:

1. Click on **Get Data** and select the **Text/CSV** connector.

2. Browse to your local folder where you downloaded the FactInternetSales CSV file and open it. A window with a preview of the data will pop up. Click on **Transform Data**.

3. Browse to the **Transform** tab and click on **Group By**:

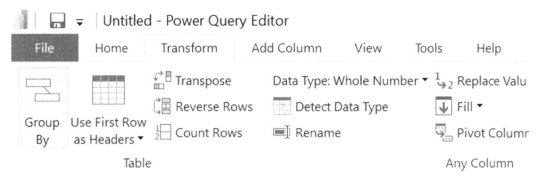

Figure 4.21 – Group By

4. Flag **Basic** and select the ProductKey column. In **New column name**, type SalesAmount, define **Sum** as **Operation** and SalesAmount as the **Column** on which to perform the sum, and click **OK**:

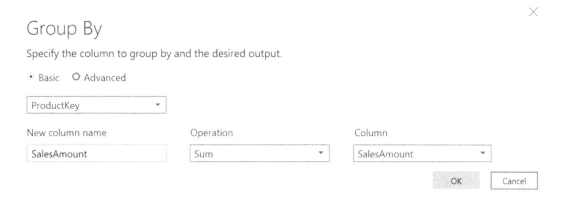

Figure 4.22 – Group By Basic

5. You can observe how the aggregation was performed by summing `SalesAmount` for each `ProductKey`:

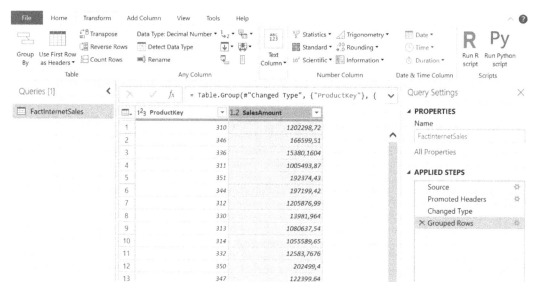

Figure 4.23 – Group By output

This is a simple aggregation based on one column, but this feature allows you to apply advanced grouping logic, as shown in the following steps:

1. Double-click on the **Grouped Rows** step in order to open the **Group By** window and edit the step we defined previously:

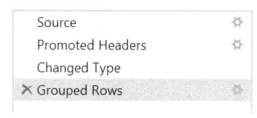

Figure 4.24 – Grouped Rows step

2. Flag **Advanced** and enter the information as seen in the following screenshot:

Group By

Specify the columns to group by and one or more outputs.

Basic ⦿ Advanced

| ProductKey | ▾ |

| SalesTerritoryKey | ▾ |

Add grouping

New column name	Operation		Column	
SalesAmount	Sum	▾	SalesAmount	▾
SalesAmountAVG	Average	▾	SalesAmount	▾
TotalCost	Sum	▾	TotalProductCost	▾
TotalCostAVG	Average	▾	TotalProductCost	▾

Add aggregation

OK   Cancel

Figure 4.25 – Group By Advanced

3. You can see that we defined an advanced grouping logic that aggregates data by `ProductKey` and `TerritoryKey`.

You have many possibilities on how to aggregate data. You can perform different built-in calculations, for example:

Figure 4.26 – Group By calculations

By leveraging these options, it is easy to simplify data structure and build aggregation tables ready to be used for the next steps.

Unpivoting and pivoting columns

You may need to change how data is displayed and turn a selected row into a column or vice versa. You may also need to create a new matrix view that can be easily used while developing reporting and custom views on data. In this recipe, we will see how to leverage pivot and unpivot features that allow us to respectively turn rows to columns or to transform columns into rows.

Getting ready

For this recipe, you need to have Power BI Desktop running on your machine. You need to download the following file in a local folder:

- `FactInternetSales` CSV file

In this example, we will refer to the `C:\Data` folder.

How to do it

Once you open your Power BI Desktop application, you are ready to perform the following steps:

1. Click on **Get Data** and select the **Text/CSV** connector.

2. Browse to your local folder where you downloaded the `FactInternetSales` CSV file and open it. A window with a preview of the data will pop up; click on **Transform Data**.

3. Click on **Choose Columns** to restrict the number of columns we are working with:

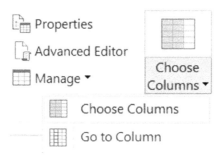

Figure 4.27 – Choose Columns

4. Select the `ProductKey`, `TotalProductCost`, `SalesAmount`, and `TaxAmt` columns and click on **OK**.

5. Select the `TotalProductCost`, `SalesAmount`, and `TaxAmt` columns by pressing *Ctrl* on your keyboard and clicking on the columns, browse to the **Transform** tab, and click on **Unpivot Columns** and then on **Unpivot Only Selected Columns**:

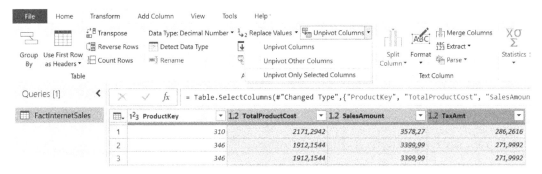

Figure 4.28 – Unpivot Columns

6. You can see that two columns were produced:

    a) `Attribute`: A column with information contained before in headers

    b) `Value`: Values that were under the previous column headers

    In the following screenshot, you can see the two new added columns and the steps applied:

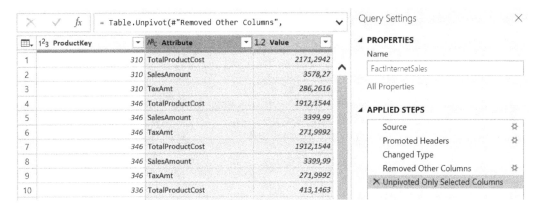

Figure 4.29 – Unpivoted values

We also have the possibility to perform the opposite. We can transform columns into rows as in the following example:

1. Remove the **Unpivoted Only Selected Columns** step from the **APPLIED STEPS** pane:

Figure 4.30 – Remove step

2. Double-click on **Removed Other Columns** and select the ProductKey, SalesAmount, and OrderDate columns:

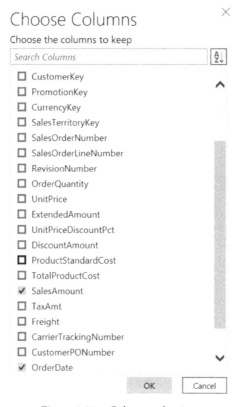

Figure 4.31 – Column selection

3. Change type to **Date** for the `OrderDate` column:

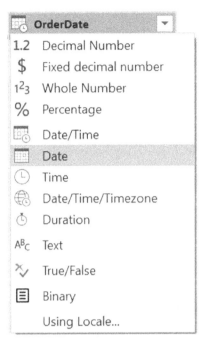

Figure 4.32 – Change data type

4. Select the `OrderDate` column, then click on **Pivot Column**:

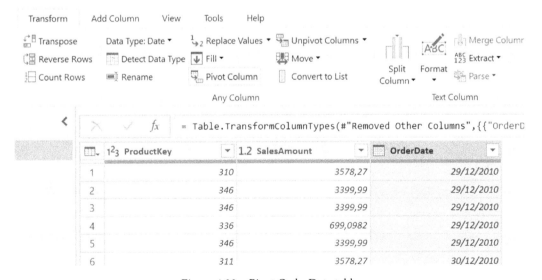

Figure 4.33 – Pivot OrderDate table

5. The **Pivot Column** window will pop up. Set **Values Column** as SalesAmount and **Aggregated Value Function** as **Sum**, and click on **OK**:

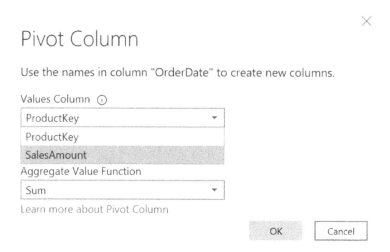

Figure 4.34 – Pivot Column window

6. You can observe how the values from the previously selected column are now column headers and the SalesAmount column values are now summed according to a *by date* logic:

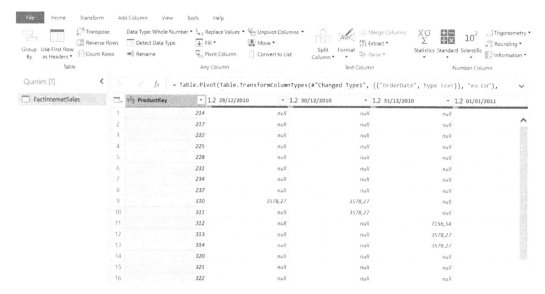

Figure 4.35 – Pivoted values

You have different options on how to aggregate data when doing the pivot transformation, as follows:

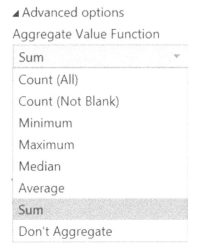

Figure 4.36 – Pivot options

For values that can't be aggregated, there is also a **Don't Aggregate** option.

In general, you can test the unpivot and pivot features and see what logic best suits your data cleaning needs.

Filling empty rows

Data sources can have their own specific structure on how data is shown. This may end up with having null values displayed on the table once this is imported into Power Query. In the following section, we will see an example that shows how you can fill missing data by keeping the original data source logic.

Getting ready

In this recipe, you need to download the following file in a local folder:

- `FactInternetSales2` CSV file

In this example, we will refer to the `C:\Data` folder.

How to do it

Once you open your Power BI Desktop application, you are ready to perform the following steps:

1. Click on **Get Data** and select the **Text/CSV** connector.

2. Browse to your local folder where you downloaded the `FactInternetSales2` CSV file and open it. A window with a preview of the data will pop up; click on **Transform Data**.

3. You can see how at `OrderDate` column level, there are some missing values:

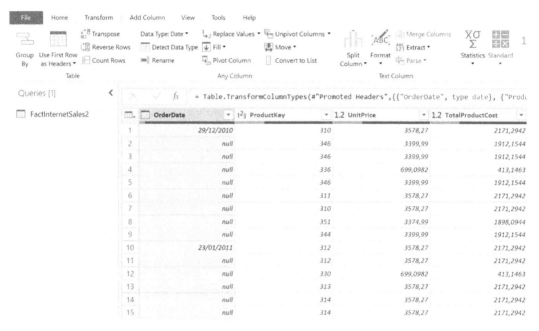

Figure 4.37 – Missing values

4. Select the `OrderDate` column, browse to the **Transform** tab, click on **Fill**, and then on **Down**:

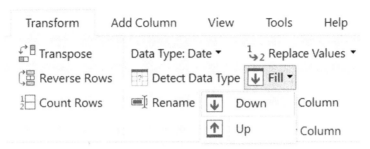

Figure 4.38 – Fill down option

5. You can see how there are no more null values, and they were filled according to existing dates:

	OrderDate	123 ProductKey	1.2 UnitPrice
1	29/12/2010	310	
2	29/12/2010	346	
3	29/12/2010	346	
4	29/12/2010	336	
5	29/12/2010	346	
6	29/12/2010	311	
7	29/12/2010	310	
8	29/12/2010	351	
9	29/12/2010	344	
10	23/01/2011	312	
11	23/01/2011	312	
12	23/01/2011	330	
13	23/01/2011	313	
14	23/01/2011	314	
15	23/01/2011	314	

Figure 4.39 – Filled values

The alternative is to fill up, if the logic behind your data requires it. In this way, you can correct your datasets in order to prepare yourself for the next use.

Splitting columns

Often, different information is merged into one column and we need to define rules to split columns and separate the information. This recipe shows how you can split data by defining custom logic according to requirements.

Getting ready

For this recipe, you need to have Power BI Desktop running on your machine. You need to download the following file in a local folder:

- `FactInternetSales` CSV file

In this example, we will refer to the `C:\Data` folder.

How to do it

Once you open your Power BI Desktop application, you are ready to perform the following steps:

1. Click on **Get Data** and select the **Text/CSV** connector.

2. Browse to your local folder where you downloaded the `FactInternetSales` CSV file and open it. A window with a preview of the data will pop up; click on **Transform Data**.

3. Browse to the `OrderDate` column and select it. Browse then to the **Transform** tab, click on **Split Column**, and then on **By Delimiter** as shown in the following screenshot:

Figure 4.40 – Split columns by delimiter

4. The **Split Column by Delimiter** window will appear. Select **Space** as the delimiter from the drop-down list, flag **Each occurrence of the delimiter**, expand **Advanced options**, flag **Columns**, check that 2 is the number of columns, and then click on **OK**:

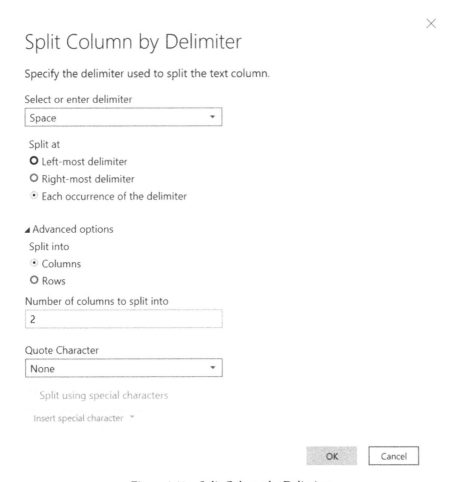

Figure 4.41 – Split Column by Delimiter

5. You will see that you end up with two columns, one with date values and the other with time values:

OrderDate.1	OrderDate.2
29/12/2010	00:00:00
29/12/2010	00:00:00
29/12/2010	00:00:00
29/12/2010	00:00:00
29/12/2010	00:00:00
30/12/2010	00:00:00
30/12/2010	00:00:00
30/12/2010	00:00:00
30/12/2010	00:00:00
31/12/2010	00:00:00
31/12/2010	00:00:00
31/12/2010	00:00:00
31/12/2010	00:00:00
31/12/2010	00:00:00

Figure 4.42 – Split columns output

There are other criteria to split columns. One of them is to split columns by the number of characters. Follow the next example to see how it works:

1. Go to the SalesOrderNumber column, click on **Split Column**, and then on **By Number of Characters**:

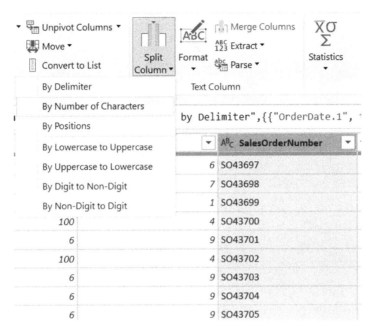

Figure 4.43 – Split Column

2. The **Split Column by Number of Characters** window will pop up. Enter 2 as **Number of characters**, flag **Once, as far left as possible**, expand **Advanced options**, flag **Columns**, and click on **OK**:

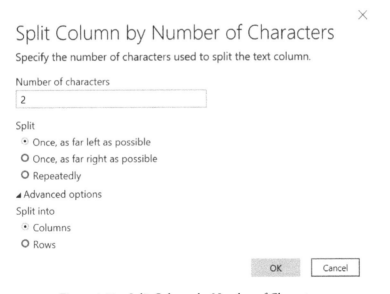

Figure 4.44 – Split Column by Number of Characters

The idea is to split the value SO from the numeric value 43697 (if we consider the values of the first row).

3. You will observe two new columns as the output of the splitting:

A^BC SalesOrderNumber.1	1²3 SalesOrderNumber.2
SO	43697
SO	43698
SO	43699
SO	43700
SO	43701
SO	43702
SO	43703
SO	43704
SO	43705
SO	43706
SO	43707
SO	43708

Figure 4.45 – Split Column output

Other methods to split columns include the following:

- By positions
- By lowercase to uppercase
- By uppercase to lowercase
- By digit to non-digit
- By non-digit to digit

You can define custom logic on how to split the data and leverage the main benefit on the Power Query side, which is to apply this step also to new appended data when refreshing the data source without the need to make this transformation at the data source level.

Extracting data

Similar to the previous recipe, you can extract subsets of data and information from columns in this recipe. In this recipe, we will see how we can easily extract information such as length, a selection of characters, or a range of data within the column. The idea is to show you how easy it is to perform these transformations quickly and intuitively.

Getting ready

For this recipe, you need to have Power BI Desktop running on your machine. You need to download the following file in a local folder:

- `FactInternetSales` CSV file

In this example, we will refer to the `C:\Data` folder.

How to do it

Once you open your Power BI Desktop application, you are ready to perform the following steps:

1. Click on **Get Data** and select the **Text/CSV** connector.

2. Browse to your local folder where you downloaded the `FactInternetSales` CSV file and open it. A window with a preview of the data will pop up. Click on **Transform Data**.

3. Browse to the **Transform** tab, click on **Extract**, and click on **Last Characters**:

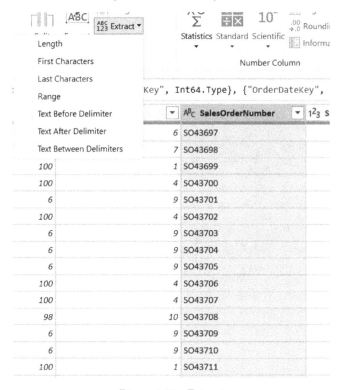

Figure 4.46 – Extract

4. The **Extract Last Characters** window will pop up, where you can enter how many characters to extract, starting from the last. In this case, enter 5 in order to extract the characters containing numbers and click on **OK**:

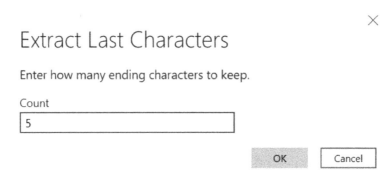

Figure 4.47 – Extract Last Characters

5. As an output, you will see that the same column was transformed:

Figure 4.48 – Extract output

Other information that can be extracted is as follows:

- Length of characters
- Defined ranges
- Text before/after delimiters
- Text between delimiters

In this way, we can keep only relevant information and rename columns with a more suitable name.

Parsing JSON or XML

We may sometimes find mixed data structures within the same query. In this recipe, we will see how to deal with mixed data structures when single columns within a table contain JSON data structure. The same reasoning would apply to XML structure too.

Getting ready

For this recipe, you need to have Power BI Desktop running on your machine. You need to download the following file in a local folder:

- `InternetSales` CSV file

In this example, we will refer to the `C:\Data` folder.

How to do it

Once you open your Power BI Desktop application, you are ready to perform the following steps:

1. Click on **Get Data** and select the **Text/CSV** connector.

2. Browse to your local folder where you downloaded the `InternetSales` CSV file and open it. The following window with a preview of the data will pop up. Click on **Transform Data**:

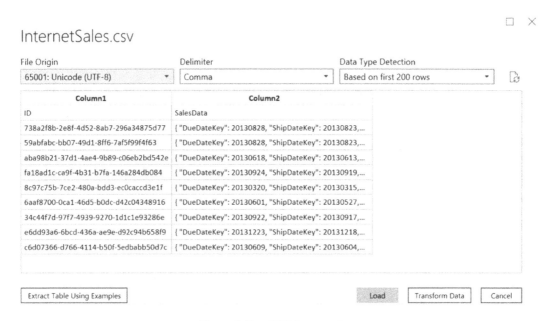

Figure 4.49 – CSV data preview

3. Browse to the **Home** tab and click on **Use First Row as Headers** and observe the data structure as follows:

    a) `ID` column

    b) `SalesData` column containing a JSON data structure with information on sales transactions

You will see the two columns displayed as in the following screenshot:

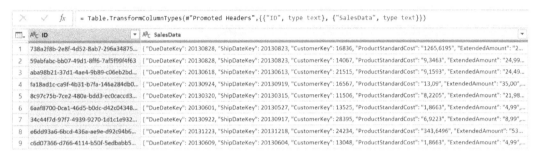

Figure 4.50 – Mixed data structure

4. Select the `SalesData` column, browse to the **Transform** tab, click on **Parse**, and then on **JSON**:

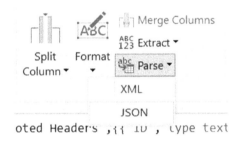

Figure 4.51 – Parse JSON

5. You will see that instead of the column with JSON data, you now have a list of records that you can expand. You can visualize and select the values that were parsed from the JSON file:

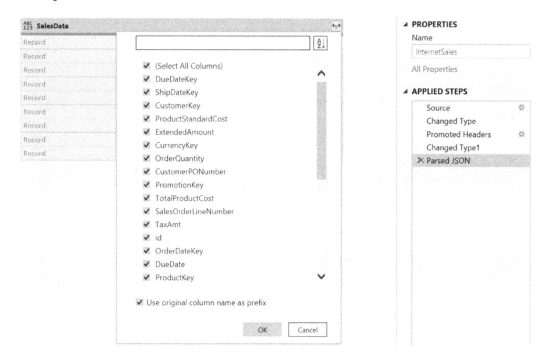

Figure 4.52 – Manage parsed data

6. Select all data and click on **OK**:

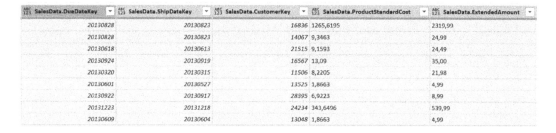

Figure 4.53 – Expanded parsed data

You can see how, with a few easy steps, you can manipulate a data structure that would be difficult to manage with other tools. This means that end users can push themselves to do advanced data preparation without the need to request changes at data source level.

The logic on JSON parsing applies to the XML structure as well.

Exploring artificial intelligence insights

Power Query allows us to enrich data with a data engineering approach. It also provides us with **artificial intelligence (AI)** tools to access **cognitive services**, which are pre-trained **machine learning (ML)**models provided by Microsoft, usually used by data scientists and app developers to apply cognitive capabilities to reading and interpreting data.

In this recipe, you will explore how to leverage these features in order to run, with a no-code approach, consistent text analysis, thanks to the use of cognitive services.

Getting ready

For this recipe, you need to have Power BI Desktop running on your machine. You need to download the following file in a local folder:

- `IMDB-Dataset` CSV file with movie reviews data

In this example, we will refer to the `C:\Data` folder.

In order to access cognitive services resources, you need to have a running Power BI Premium capacity.

How to do it

Once you open your Power BI Desktop application, you are ready to perform the following steps:

1. Click on **Get Data** and select the **Text/CSV** connector.

2. Browse to your local folder where you downloaded the IMDB-Dataset CSV file and open it. The following window with a preview of the data will pop up. Click on **Transform Data**:

Figure 4.54 – CSV data preview

3. Browse to the **Home** tab and click on **Use First Rows as Headers**. Select the first column, review, where you can find movie reviews, browse to the **Add Column** tab, and click on **Text Analytics** in order to see the following window displayed:

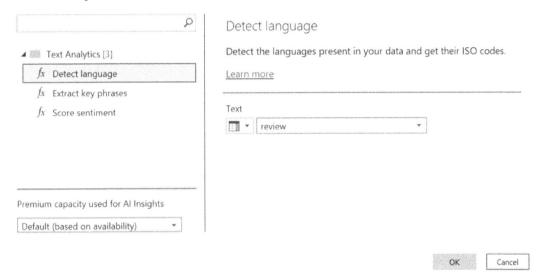

Figure 4.55 – Detect language

Now, click on **Detect language**, select the review column and click on **OK**. This way, you will be applying the text analytics function to detect the language type of the selected column.

4. You will see two new columns, Detect language.Detected Language Name and Detect language.Detected Language ISO Code, with information about the language detected:

ABC Detect language.Detected Language Name	ABC Detect language.Detected Language ISO Code
English	en
English	en
English	en
English	en
English	en

Figure 4.56 – Detect language output

5. Click again on **Text Analytics** and click now on **Extract key phrases**:

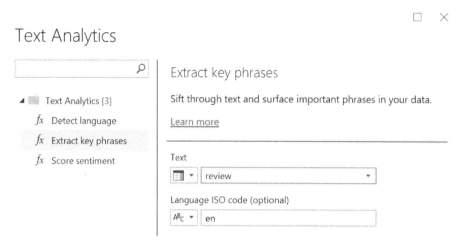

Figure 4.57 – Extract key phrases

6. You can see the following two columns with key words for every single review:

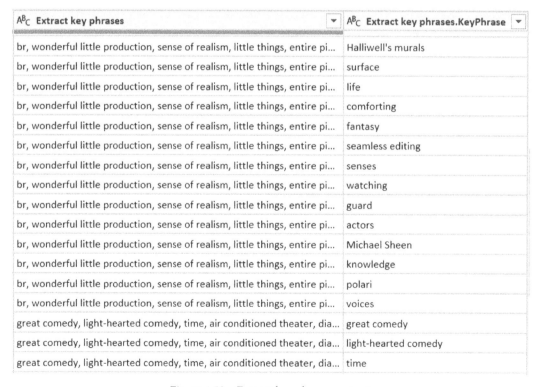

ABC Extract key phrases	ABC Extract key phrases.KeyPhrase
br, wonderful little production, sense of realism, little things, entire pi...	Halliwell's murals
br, wonderful little production, sense of realism, little things, entire pi...	surface
br, wonderful little production, sense of realism, little things, entire pi...	life
br, wonderful little production, sense of realism, little things, entire pi...	comforting
br, wonderful little production, sense of realism, little things, entire pi...	fantasy
br, wonderful little production, sense of realism, little things, entire pi...	seamless editing
br, wonderful little production, sense of realism, little things, entire pi...	senses
br, wonderful little production, sense of realism, little things, entire pi...	watching
br, wonderful little production, sense of realism, little things, entire pi...	guard
br, wonderful little production, sense of realism, little things, entire pi...	actors
br, wonderful little production, sense of realism, little things, entire pi...	Michael Sheen
br, wonderful little production, sense of realism, little things, entire pi...	knowledge
br, wonderful little production, sense of realism, little things, entire pi...	polari
br, wonderful little production, sense of realism, little things, entire pi...	voices
great comedy, light-hearted comedy, time, air conditioned theater, dia...	great comedy
great comedy, light-hearted comedy, time, air conditioned theater, dia...	light-hearted comedy
great comedy, light-hearted comedy, time, air conditioned theater, dia...	time

Figure 4.58 – Extract key phrases output

As well as text analytics services, you can also do analysis on images using vision services. If you have a column with links redirecting to images, you can enrich that content by applying image tags.

Moreover, you can recall custom machine learning models developed on Azure Machine Learning by data scientists. The idea is to have Power Query features that allow business users and analysts to collaborate with more technical users, such as data engineers and data scientists.

Artificial intelligence insights can be seen as the entry point to recall pre-calculated artificial intelligence services such as cognitive services and custom ML models developed on Azure Machine Learning.

5
Combining Queries for Efficiency

Business analysts need to perform complex transformations that usually involve a combination of multiple queries. They often need to join data horizontally or to append tables.

With different ways of combining data, you can transform and model tables in order to optimize the information included. By leveraging different methods, you can perform merge and join transformations in order to enrich data or append and combine queries to scale and increase data volume automatically. The main aim is to create queries with relevant data that can serve different purposes, such as reporting, loading to **Dataverse** through **Power Apps**, or loading to **Azure Data Lake** to make this data available for other applications.

In this chapter, you will explore the following combining options within **Power Query**:

- Merging queries
- Joining methods
- Appending queries
- Combining multiple files
- Using the Query Dependencies view

Technical requirements

For this chapter, you will be using Power BI Desktop (`https://www.microsoft.com/en-us/download/details.aspx?id=58494`).

The minimum requirements for installation are listed here:

- .NET Framework 4.6 (Gateway release August 2019 and earlier)

- .NET Framework 4.7.2 (Gateway release September 2019 and later)

- A 64-bit version of Windows 8 or a 64-bit version of Windows Server 2012 R2 with current **Transport Layer Security** (**TLS**) 1.2 and cipher suites

- 4 **gigabytes** (**GB**) disk space for performance monitoring logs

You can find the data resources referred to in this chapter at `https://github.com/PacktPublishing/Power-Query-Cookbook/tree/main/Chapter05`.

Merging queries

Users usually need to merge data horizontally and enrich a table with additional columns that are not available within the main query when it is loaded from a data source.

In this recipe, you will see how to perform this merging and which steps to consider in order to get a successful result.

Getting ready

For this recipe, you need to download the following files:

- `FactInternetSales` CSV file

- `DimTerritory` CSV file

In this example, we will refer to the `C:\Data` folder.

How to do it...

Once you open your Power BI Desktop application, you are ready to perform the following steps:

1. Click on **Get Data** and select the **Text/CSV** connector.

2. Browse to your local folder where you downloaded the `FactInternetSales` CSV file and open it. The following window with a preview of the data will pop up; click on **Transform Data**:

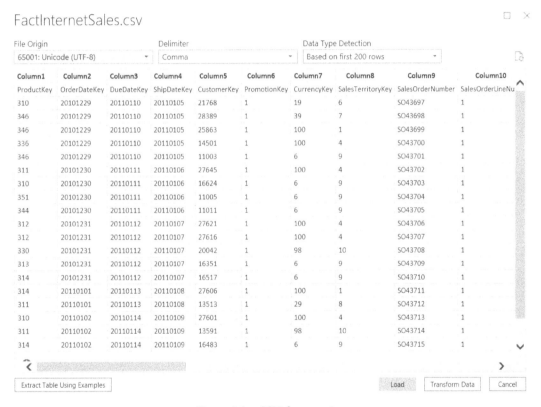

Figure 5.1 – CSV data preview

3. Repeat these steps for the `DimTerritory` CSV file in order to end up with two queries in the Power Query **user interface** (**UI**)—`FactInternetSales` and `DimTerritory`:

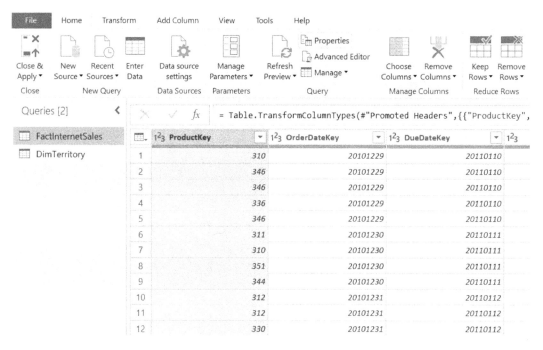

Figure 5.2 – Power Query UI

4. In this case, we want to enrich the `FactInternetSales` table with some columns coming from the `DimTerritory` table in order to get details of the geographical location of sales transactions. For this, you need to browse to the end of the **Home** tab and click on **Combine** and then **Merge Queries**:

Figure 5.3 – Merge Queries button

5. A **Merge** window will pop up. Select the `SalesTerritoryKey` column and select the `DimTerritory` table from the drop-down menu, as shown in the following screenshot:

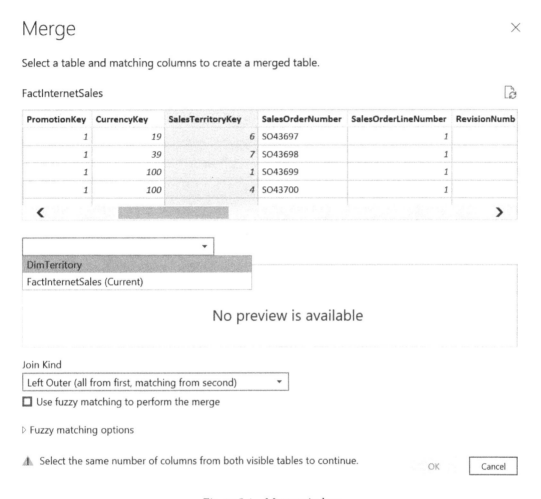

Figure 5.4 – Merge window

6. In the second table, select the `SalesTerritoryKey` column. Select **Left Outer (all from first, matching from second)** for the **Join Kind** field. Leave the other options as they are and click on **OK**:

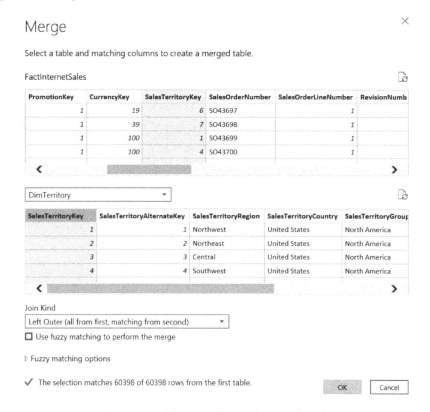

Figure 5.5 – Merge window columns selected

7. After you click **OK**, you will see that a new column will be added to the `FactInternetSales` query:

Figure 5.6 – New column added

8. Click on the **Expand** icon on the right of the DimTerritory column, select
    SalesTerritoryRegion and SalesTerritoryCountry, and click on **OK**:

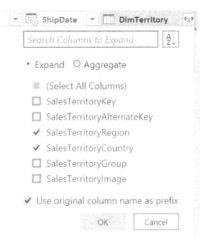

Figure 5.7 – New columns selection

9. You will see two new columns coming from the other query (DimTerritory),
    matching the rows of the main table:

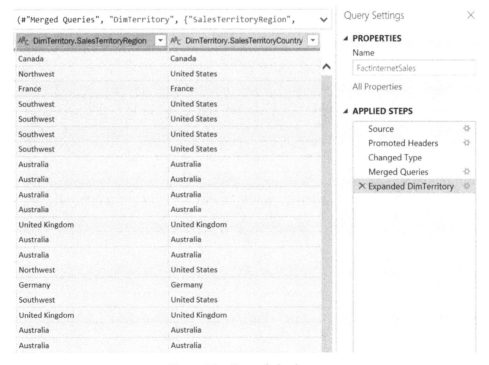

Figure 5.8 – Expanded columns

You can see how this feature allows users to enrich and transform data coming from different sources, without the need to make any change to the data source directly.

Joining methods

In the previous recipe, you had the chance to see how to perform a merge where you reference a main table enriched with data coming from another table with geographical details. In fact, there are many ways to join data on matching values following a logic that belongs to traditional relational databases—for example, left/right/full outer joins, inner joins, and left/right anti-joins. These different methods allow users to match data by applying custom logic.

In this recipe, you will see how you can effectively leverage some of the most popular joining methods.

Getting ready

For this recipe, you need to download the following files:

- `FactInternetSales` CSV file
- `DimTerritory2` CSV file

In this example, we will refer to the `C:\Data` folder.

How to do it…

Once you open your Power BI Desktop application, you are ready to perform the following steps:

1. Click on **Get Data** and select the **Text/CSV** connector.
2. Browse to your local folder where you downloaded the `FactInternetSales` CSV file and open it. The following window with a preview of the data will pop up; click on **Transform Data**:

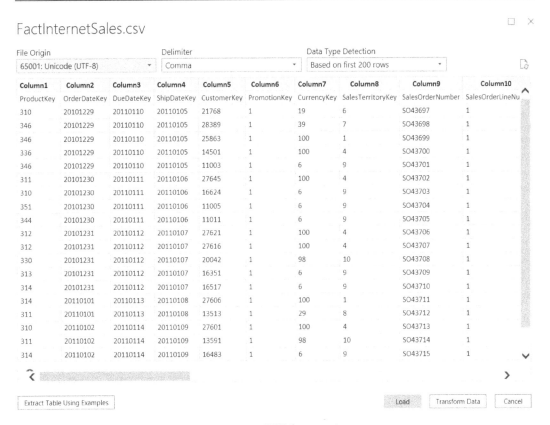

Figure 5.9 – CSV data preview

3. Repeat *Steps 1* and *2* for the `DimTerritory2` CSV file in order to end up with the following two queries in the Power Query UI:

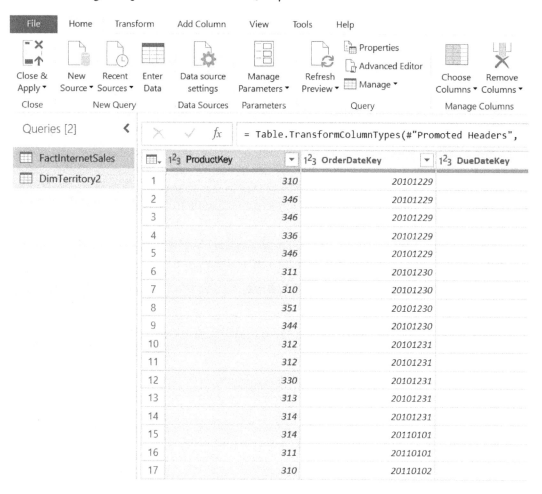

Figure 5.10 – Power Query UI

4. In this case, we want to enrich `DimTerritory2` with aggregated data coming from the `FactInternetSales` table by performing a right outer join. For this, you need to select `DimTerritory2` in the **Queries** pane, browse to the end of the **Home** tab, and click on **Merge Queries**:

Figure 5.11 – Merge Queries button

5. Select `SalesTerritoryKey` from the `DimTerritory2` query and `FactInternetSales` from the drop-down menu, selecting the same matching value. In the **Join Kind** field, select **Right Outer (all from second, matching from first)** and click on **OK**:

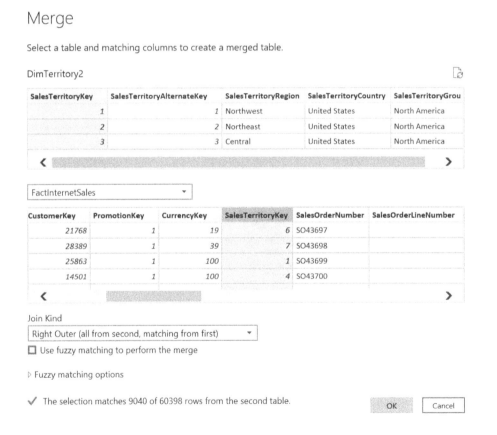

Figure 5.12 – Merge window

6. In the `DimTerritory2` query, you can find a subset of three rows we had in the original table (`DimTerritory`) from the previous recipe and a row with `null` values. When merging with a right outer join, you enriched the `DimTerritory2` table with both matching and non-matching values. Click on the **Expand** icon on the left of the `FactInternetSales` column, flag **Aggregate**, select **Sum of TotalProductCost** and **Sum of SalesAmount**, unflag **Use original column name as prefix**, and click on **OK**:

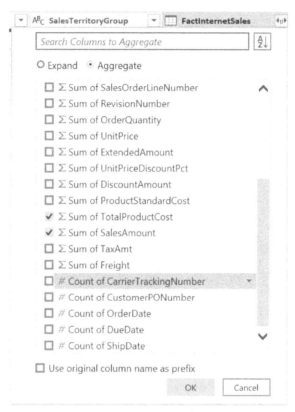

Figure 5.13 – Expanding merged column

7. You can see two new columns containing data from the `FactInternetSales` table:

ABC 123 Sum of TotalProductCost	ABC 123 Sum of SalesAmount
2130235.255	3649866.551
15142278.71	25699277.37
3629.7059	6532.4682
1649.9092	3000.8296

Figure 5.14 – Added columns

You can see that in this way, you can add aggregated data and enrich tables with external data. This is useful when you need both matching and non-matching data. In this recipe, you added aggregated sales and total cost values for the geographies available in the `DimTerritory2` table, and you also added a row that does not have a match to represent sales and total costs for geographies that are not mapped.

In the previous recipe, you saw how to enrich data coming from other queries. In some cases, you may need to enrich data only on matching values, as you will see in the following exercise.

If you double-click on the **Merged Queries** step on the **APPLIED STEPS** pane, you can change and explore other join possibilities. Follow the next steps to see how:

1. Edit the **Merge Queries** step by double-clicking on it, wait for the **Merge** window to appear, change **Join Kind** to **Inner (only matching rows)**, and click on **OK**:

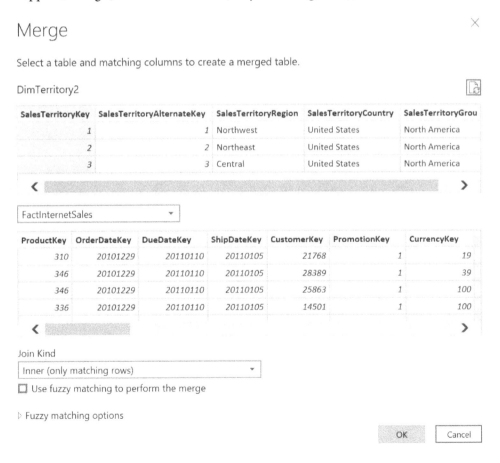

Figure 5.15 – Merge window

2. You can see that in this way, we only added matching information:

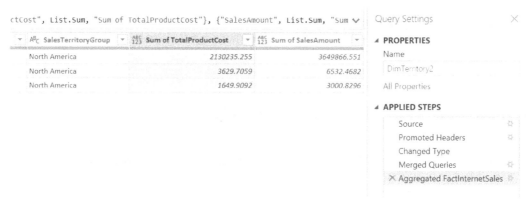

Figure 5.16 – Inner join output

Let's add another query to test another join method, **Left Anti**:

1. Click on **Get Data** and select the **Text/CSV** connector.

2. Browse to your local folder where you downloaded the DimProduct CSV file and open it. A preview of the data will pop up; click on **Transform Data**.

3. Select DimProduct from the **Queries** pane, browse to the **Home** tab, and click on **Merge queries**.

4. Select ProductKey from the DimProduct query and FactInternetSales from the drop-down menu, selecting the same matching value. In the **Join Kind** field, select **Left Anti (rows only in first)** and click on **OK**:

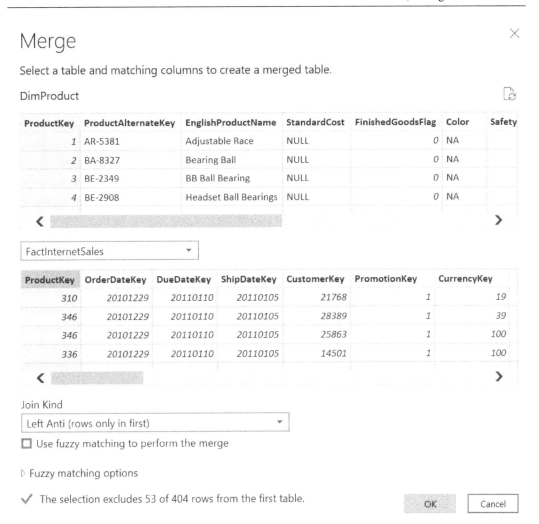

Figure 5.17 – Merge window

The idea with this join kind is to have a dataset with all rows from the first table less the matching rows from the second table. If the DimProduct table originally has 404 rows, after this join it will have 53 fewer rows, which are the rows that are matching from FactInternetSales.

5. Click on the **Expand** icon on the left of the Fact InternetSales column, flag **Expand**, select ProductKey, and click on **OK**:

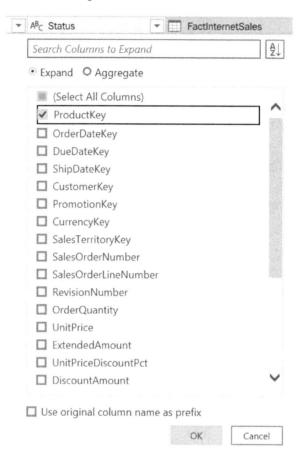

Figure 5.18 – Expanding merged column

6. You can see that the number of rows is reduced in the bottom left of the page:

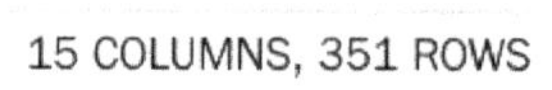

15 COLUMNS, 351 ROWS

Figure 5.19 – Updated number of rows

As we observed in this recipe, there are many ways to perform merge transformations, and each allows us to get different outputs of data. You can enrich data, reduce it, retrieve values for each row, or perform built-in aggregations. This step allows you to easily edit update join transformations and tells you which data to expand.

Appending queries

Within a single table, you often need to have data coming from different files/tables. You need to append data and have a unique view that will allow you to run more complex analyses. In this recipe, you will see how you can append data in Power Query with just a few clicks.

Getting ready

For this recipe, you need to download the following files:

- `FactInternetSales` CSV file
- `FactResellerSales` CSV file

In this example, we will refer to the `C:\Data` folder.

How to do it...

Once you open your Power BI Desktop application, you are ready to perform the following steps:

1. Click on **Get Data** and select the **Text/CSV** connector.
2. Browse to your local folder where you downloaded the two CSV files (`FactInternetSales` and `FactResellerSales`) and load them into the Power Query view:

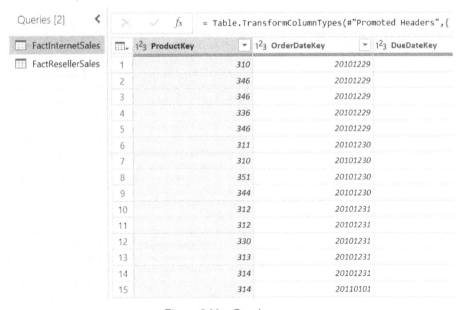

Figure 5.20 – Queries pane

3. Select the `FactInternetSales` query, browse to the **Add Column** tab, and click on **Custom Column**:

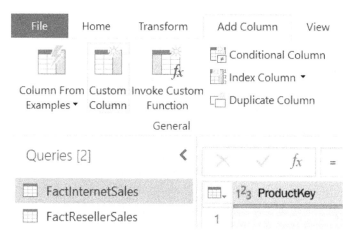

Figure 5.21 – Custom Column button

4. Create a new column called `Channel`, add as a formula the value `Internet`, and click on **OK**:

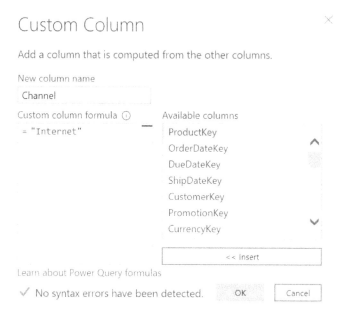

Figure 5.22 – Custom Column window

Repeat this step for the `FactResellerSales` query, with the only difference being on the value to define in the added column. Instead of `Internet`, enter `Reseller`:

Custom Column   ✕

Add a column that is computed from the other columns.

New column name

Channel

Custom column formula ⓘ

```
= "Reseller"
```

Available columns

ProductKey
OrderDateKey
DueDateKey
ShipDateKey
CustomerKey
PromotionKey
CurrencyKey

<< Insert

Learn about Power Query formulas

✓ No syntax errors have been detected.   OK   Cancel

Figure 5.23 – Custom Column window: Reseller

You should end up with an added column for each query, as shown in the following screenshot:

Figure 5.24 – New added columns

5. Browse to the **Home** tab and click on **Append Queries as New**:

Figure 5.25 – Append Queries as New button

6. Select FactInternetSales in the **First table** field and FactResellerSales in the **Second table** field, and click on **OK**:

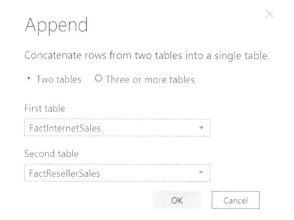

Figure 5.26 – Append window

7. You will end up with a query named Append1 (you can rename this as you wish), with appended data from the two queries. In this case, some column headers do not match between the two tables, such as CustomerKey from FactInternetSales and ResellerKey from FactResellerSales. In this case, the Append1 query will show null values:

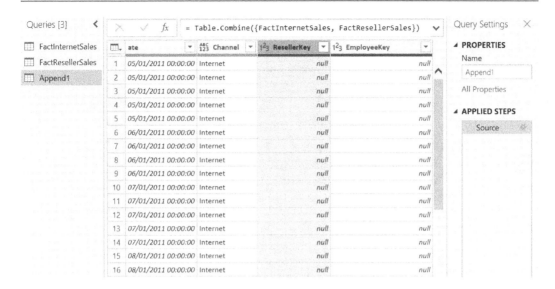

Figure 5.27 – Null values for non-matching columns

You can append more than two tables and apply the same logic to a wider number of queries.

Combining multiple files

While working with your data, you may need to automatically combine multiple files. You could use the append transformation method, but if you imagine a use case where there are files loaded into a folder with a defined frequency and you need to see the new data coming at each refresh, it is clear that an alternative method is needed. In this recipe, you will see how to connect and combine multiple files with just a few clicks, regardless of how many there are in the folder.

Getting ready

For this recipe, to test different types of file connectors, you need to download a `CSVFiles` folder containing CSV files, as shown in the following screenshot:

InternetSales20110123	Microsoft Excel Comma Separated Values File	2 KB
InternetSales20110124	Microsoft Excel Comma Separated Values File	3 KB
InternetSales20110125	Microsoft Excel Comma Separated Values File	3 KB
InternetSales20110126	Microsoft Excel Comma Separated Values File	2 KB
InternetSales20110127	Microsoft Excel Comma Separated Values File	2 KB
InternetSales20110128	Microsoft Excel Comma Separated Values File	2 KB
InternetSales20110129	Microsoft Excel Comma Separated Values File	2 KB
InternetSales20110130	Microsoft Excel Comma Separated Values File	1 KB
InternetSales20110131	Microsoft Excel Comma Separated Values File	2 KB
InternetSales20110201	Microsoft Excel Comma Separated Values File	2 KB

Figure 5.28 – Local folder with CSV files

In this example, I will refer to the following path: `C:\Data\CSVFiles`.

How to do it...

Open the Power BI Desktop application to perform the following steps:

1. Go to **Get Data**, then click on the **Folder** connector. You can directly enter your folder path or click on **Browse** and select the folder from the usual browsing section of your machine. Then, click on **OK**.

2. You will see the following window with a list of files contained in the folder:

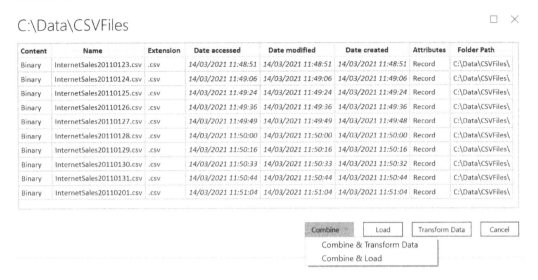

Figure 5.29 – How files from the folder are displayed

3.   Click on **Transform Data** and you will see the following columns:

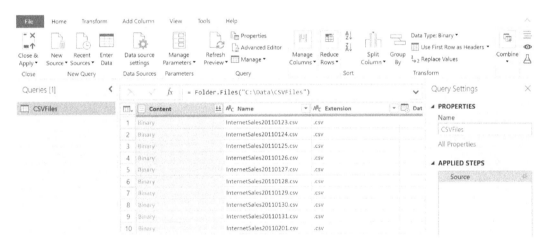

Figure 5.30 – List of files in Power Query view

4. Click on the **Combine** icon, which you will find in the left corner of the `Content` column, as shown in the following screenshot:

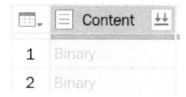

Figure 5.31 – Combine icon

5. The **Combine Files** window will pop up. From here, you can define which file to use in the **Sample File** field and define the **File Origin**, **Delimiter**, and **Data Type Detection** fields. Leave the detected values you see for each field and click on **OK**, as seen in the following screenshot:

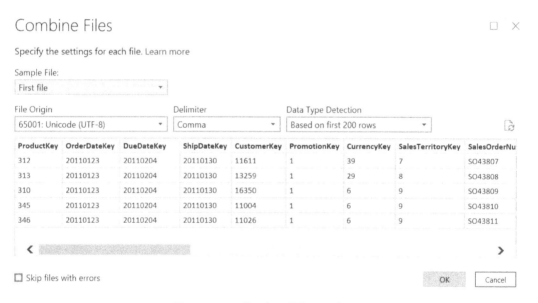

ProductKey	OrderDateKey	DueDateKey	ShipDateKey	CustomerKey	PromotionKey	CurrencyKey	SalesTerritoryKey	SalesOrderNu
312	20110123	20110204	20110130	11611	1	39	7	SO43807
313	20110123	20110204	20110130	13259	1	29	8	SO43808
310	20110123	20110204	20110130	16350	1	6	9	SO43809
345	20110123	20110204	20110130	11004	1	6	9	SO43810
346	20110123	20110204	20110130	11026	1	6	9	SO43811

Figure 5.32 – Combine Files window

6. You can see that all data from all files is collected in one query:

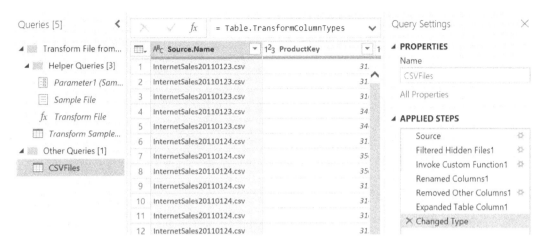

Figure 5.33 – Combine Files output

With the **Combine** transformation method, there will be a set of transformations that are automatically generated by the system. These include the definition of a sample file and a function that iterates over all the files in the folder and then expands and appends them.

Users can customize this generated function and adapt it to their custom use cases. An example will be explored in *Chapter 8, Adding Value to Your Data*.

Using the Query Dependencies view

Once you become confident using Power Query, you can start adding a higher number of queries, combining them, and applying complex transformation logic. You can often get lost in terms of how queries were built and merged and you may need a view that allows you to map data sources and queries. In this recipe, we will see how the **Query Dependencies** view will allow us to quickly see what is going on in that Power Query session.

Getting ready

In this recipe, you will need to connect to **Azure SQL Database** with AdventureWorks data. You need to have access to a running database.

How to do it...

Once you open Power BI Desktop, perform the following steps:

1. Go to **Get data**, click on **More…**, and browse for **Azure SQL database**:

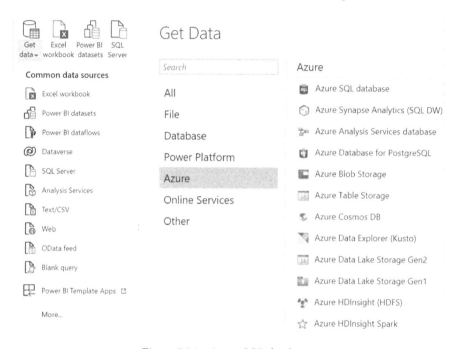

Figure 5.34 – Azure SQL database

2. Enter the server name of your Azure SQL database:

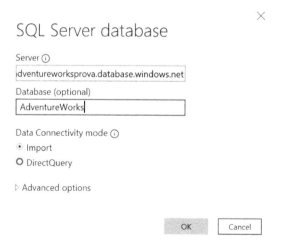

Figure 5.35 – SQL Server information

3. Enter authentication details:

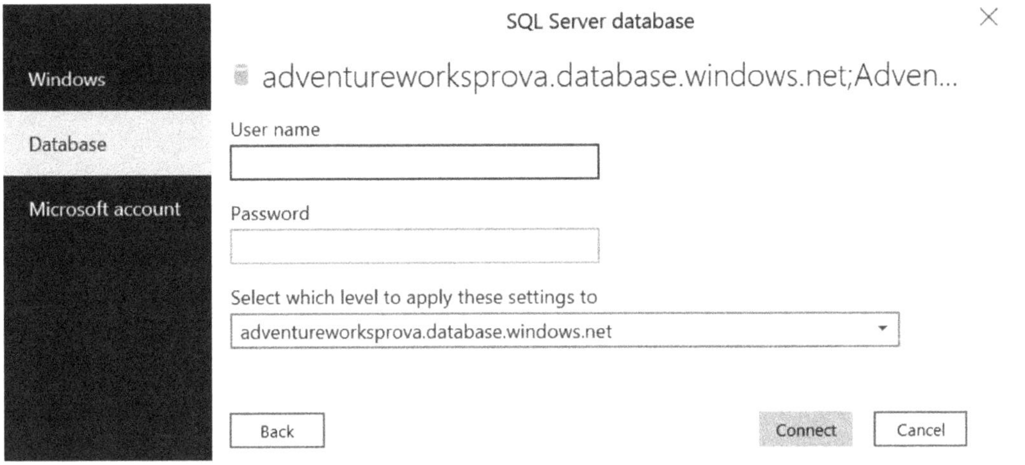

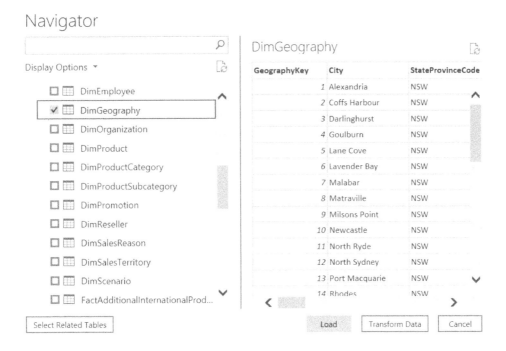

Figure 5.36 – Azure SQL database authentication

4. Flag DimGeography, FactInternetSales, and FactResellerSales, and click on **Transform Data**:

Figure 5.37 – Table selection

5. Select `FactInternetSales`, click on **Choose Columns**, type `Dim` in the search bar, and unflag all columns:

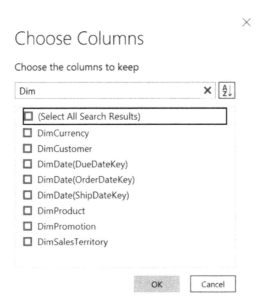

Figure 5.38 – Unflagging Dim columns

6. Click on **OK** and repeat this step for the `FactResellerSales` query.

    Select the `FactResellerSales` query and click on **Append Queries as New**:

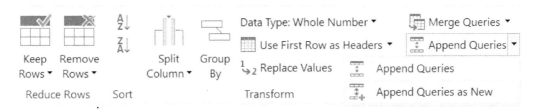

Figure 5.39 – Append Queries as New button

7. Select `FactInternetSales` in the **Second table** field:

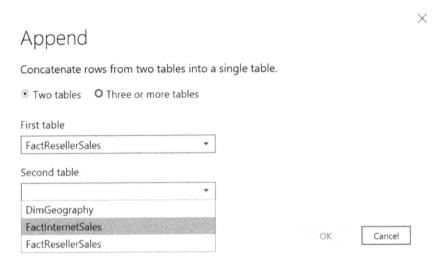

Figure 5.40 – Append Queries detail

8. Rename the newly created query as `TotalSales`:

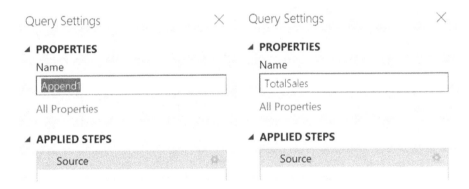

Figure 5.41 – Query Settings properties

9. Browse to the **View** tab and click on **Query Dependencies**:

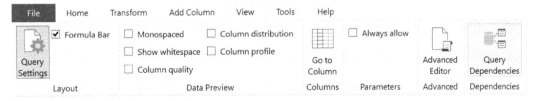

Figure 5.42 – Query Dependencies button

A tree view will open up all queries in the current Power Query session in a mapped format:

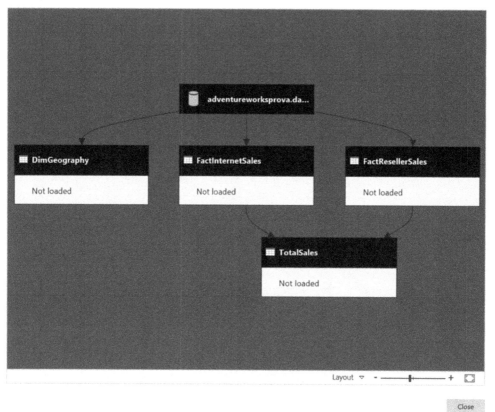

Figure 5.43 – Query Dependencies view

10. On the bottom right of the screen, click on **Layout** and click on **Left to Right Layout**:

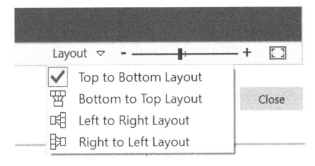

Figure 5.44 – Query Dependencies layout

11. A different layout of the view will be displayed, and in the same way, you can change it according to your preferred view. This clearly depends on the complexity and number of queries:

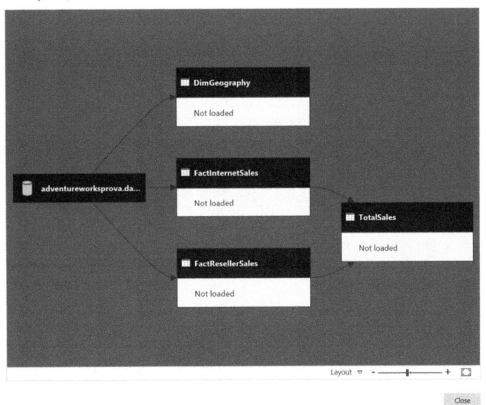

Figure 5.45 – Query Dependencies changed layout

12. The dependency tree helps users to map all queries and their dependencies. In this case, we can see how a new `TotalSales` query is generated by `FactInternetSales` and `FactResellerSales`.

It is possible to use this view to view the overall Power Query design and understand which data sources have been used and what is the level of the queries' complexity in terms of dependency.

6

Optimizing Power Query Performance

Once you become more confident with **Power Query** and are able to do the main transformations and reshape your data, you are ready to focus on optimizing queries' performance. This topic is important because when you prepare data for reporting or any other further use and you need to refresh data or to retrieve subsets of filtered data, you need to think about and design Power Query steps in a way that helps you to avoid slow queries and suboptimal performance for end users. In this chapter, we will learn how to use concepts such as parameters and query folding that can help you improve queries' loading times.

In particular, you will explore the following performance optimization options within Power Query:

- Setting up parameters
- Filtering with parameters
- Folding queries
- Leveraging incremental refresh and folding
- Disabling query load

Technical requirements

For this chapter, you will be using the following:

- **Power BI Desktop**: `https://www.microsoft.com/en-us/download/details.aspx?id=58494`

The minimum requirements for installation are the following:

- NET Framework 4.6 (Gateway release August 2019 and earlier)
- NET Framework 4.7.2 (Gateway release September 2019 and later)
- A 64-bit version of Windows 8 or a 64-bit version of Windows Server 2012 R2 with current TLS 1.2 and cipher suites
- 4 GB disk space for performance monitoring logs

You can find the data resources referred to in this chapter at `https://github.com/PacktPublishing/Power-Query-Cookbook/tree/main/Chapter06`.

Setting up parameters

You have the possibility to manage in a flexible way which data to load in the model. You can define parameters and use them in inputs for multiple transformations such as filtering or enriching data. In this recipe, we will see how to define parameters and how to use them when adding a conditional column based on some threshold values where you want to add a flag for each value of an existing column.

Getting ready

For this recipe, you need to download the `FactInternetSales` CSV file.

In this example, we will refer to the `C:\Data` folder.

How to do it...

Once you open your Power BI Desktop application, you are ready to perform the following steps:

1. Click on **Get data** and select the **Text/CSV** connector:

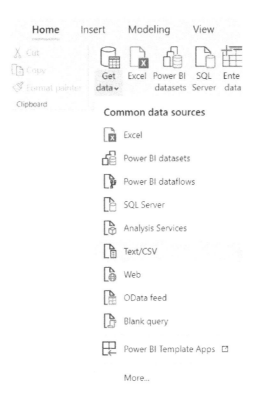

Figure 6.1 – Text/CSV connector

2. Browse to your local folder where you downloaded the `FactInternetSales`
    CSV file and open it. The following window with a preview of the data will pop up.
    Click on **Transform Data**:

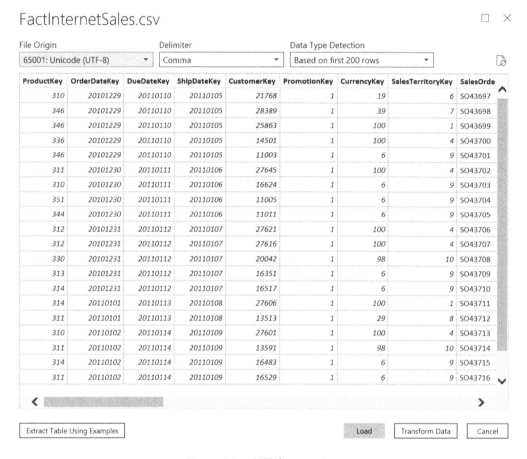

Figure 6.2 – CSV data preview

3. Browse to the **Home** tab and click on **Manage Parameters**:

Figure 6.3 – Manage Parameters button

4. Create two parameters, `Parameter1` and `Parameter2`, and enter for each the information seen in the following screenshots, and then click on **OK**:

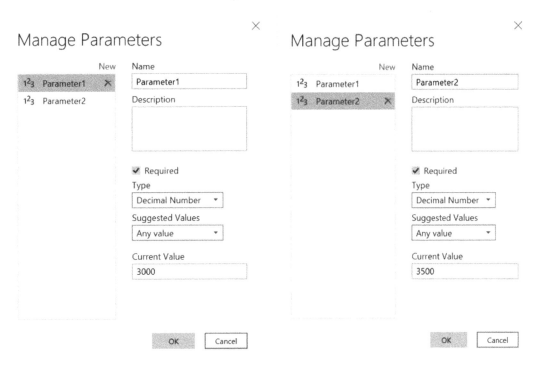

Figure 6.4 – (Left) Parameter1 creation. (Right) Parameter2 creation

5. You will see that two new elements, **Parameter1 (3000)** and **Parameter2 (3500)**, are visible in the Power Query window in the **Queries** pane.

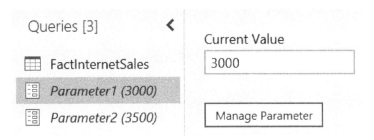

Figure 6.5 – Parameters in the Queries pane

6. Now, select the FactInternetSales query (above **Parameter 1 (3000)**), browse to the **Add Column** tab, and click on **Conditional Column**, as seen in the following screenshot:

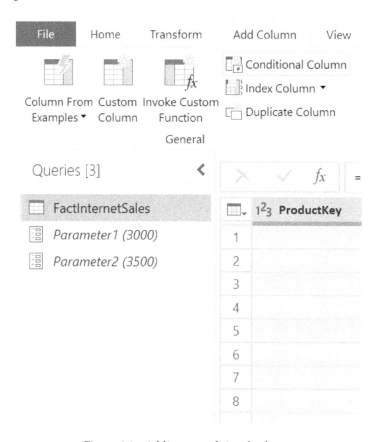

Figure 6.6 – Adding a conditional column

7. Create a conditional column, SalesLevel, and enter the values for each condition, as shown in the following screenshot, and then click on **OK**:

Add Conditional Column

×

Add a conditional column that is computed from the other columns or values.

New column name

SalesLevel

	Column Name	Operator	Value ⓘ				Output ⓘ	
If	SalesAmount ▾	is greater than ▾	▦ ▾	Parameter2 ▾	Then	ABC 123 ▾	High
Else If	SalesAmount ▾	is greater than ▾	▦ ▾	Parameter1 ▾	Then	ABC 123 ▾	Medium	⋯

Add Clause

Else ⓘ

ABC 123 ▾   Low

OK   Cancel

Figure 6.7 – Add Conditional Column window

The conditional columns allow us to define three labels, High, Medium, or Low, for each SalesAmount value depending on the values of the parameters we define.

8. You will see a new column in the FactInternetSales query, called SalesLevel, with values that reflect different labels according to the value contained in the SalesAmount column, as in the following screenshot:

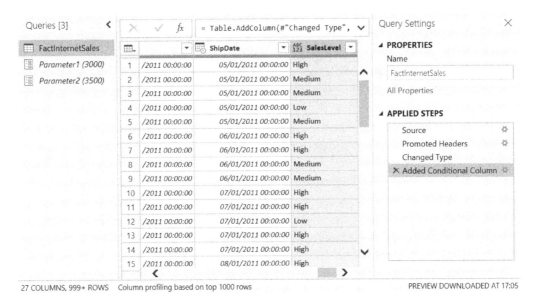

Figure 6.8 – New added column

9. Select **Parameter2** and edit the value by entering 3350, and then press the *Enter* button on your keyboard.

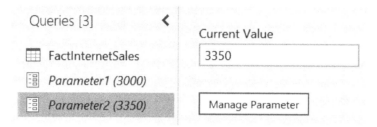

Figure 6.9 – Editing the parameter value

10. Select the FactInternetSales query and browse to the SalesLevel column that you created in the previous steps. See how the labels have changed according to the value of **Parameter2**.

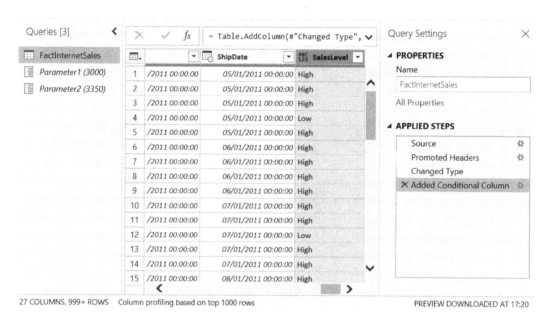

Figure 6.10 – Updated labels

Parameters are useful because, as you saw in this example, they allow us to dynamically change thresholds and to apply a transformation without the need for editing it manually with static values, but rather adapting it with parameters.

Filtering with parameters

Parameters are a key functionality when it comes to the definition of dynamic filtering logic. You can create parameters that will be used to filter and load data according to predefined values. This way, you will be able to work on a subset of data optimizing general queries' performance.

In this recipe, you will see how to create a parameter over a key value, for example, a product key.

Getting ready

For this recipe, you need to download the `FactInternetSales` CSV file.

In this example, we will refer to the `C:\Data` folder.

How to do it...

Once you open your Power BI Desktop application, you are ready to perform the following steps:

1. Click on **Get data** and select the **Text/CSV** connector.

Figure 6.11 – Text/CSV connector

2. Browse to your local folder where you downloaded the `FactInternetSales.csv` file and open it. The following window with a preview of the data will pop up. Click on **Transform Data**.

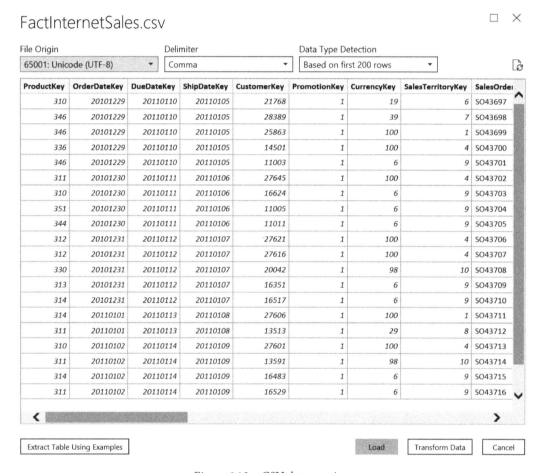

Figure 6.12 – CSV data preview

3. Browse to the **Home** tab and click on **Manage Parameters**.

Figure 6.13 – Manage Parameters button

4. Create a parameter called `ProductKey`, select **Decimal Number** for **Type** from the relative drop-down section, select **List of values** from the **Suggested Values** dropdown, and enter the values 310, 346, and 336, as shown in the following screenshot:

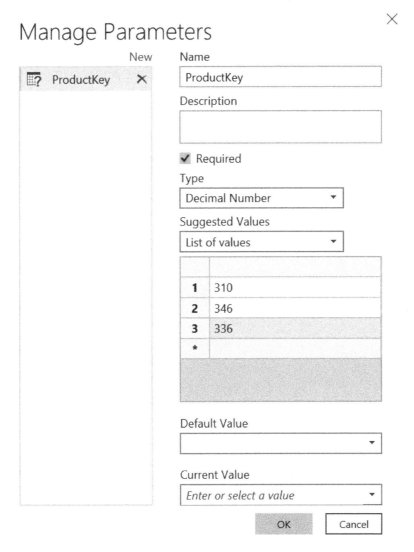

Figure 6.14 – Manage Parameters window

5. Select the value **310** from the dropdowns for both **Default Value** and **Current Value** and click on **OK**.

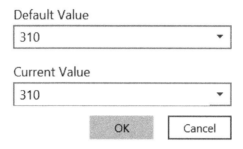

Figure 6.15 – Defining default and current values

6. Browse to the **Queries** pane, select the **ProductKey (310)** parameter, and observe how you can select one of the three values you defined for that parameter, as shown in the following screenshot:

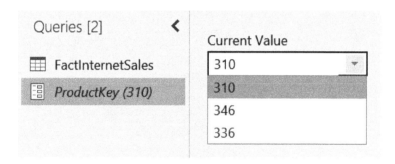

Figure 6.16 – ProductKey parameter

7. From the **Queries** pane, select the `FactInternetSales` query. Then, select the `ProductKey` column from the query selected. Then, click on the drop-down icon on the right part of the `ProductKey` column and click on **Number Filters** and then **Equals…**, as shown in the following screenshot:

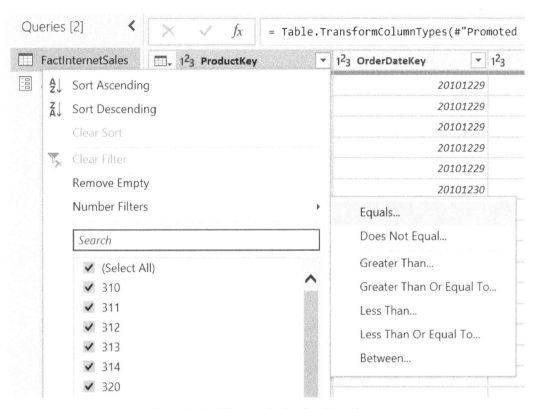

Figure 6.17 – Filtering the ProductKey column

8. Click on the type of value you want to base your equality on (the default one is **1.2**) and select **Parameter**.

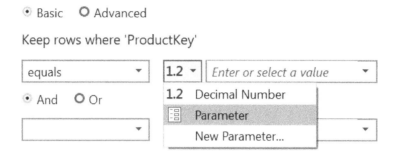

Figure 6.18 – Filter Rows window

9. **ProductKey** will be selected automatically since it is the only parameter in the current session. After this, click on **OK**.

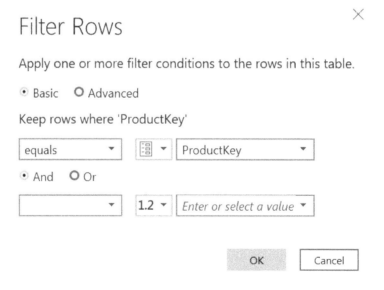

Figure 6.19 – Filtering with parameters

10. You can see how the `ProductKey` column is filtered on the current value `310` that we defined previously:

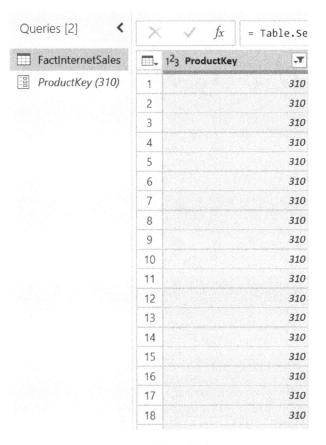

Figure 6.20 – Filtered column

11. Select the **ProductKey (310)** parameter on the **Queries** pane and change the value of the parameter from the drop-down section by selecting **346**.

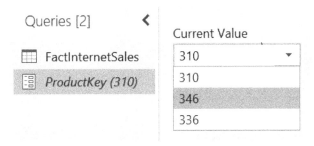

Figure 6.21 – Selecting a parameter value

12. Select the `FactInternetSales` query and observe how the filtered data changes.

Figure 6.22 – Filtered column

You can see how it is easy to dynamically apply filters with the use of parameters by defining a list of values.

In this example, we used a list of three `ProductKey` values that we created manually, but what if we want to retrieve this list from an external query?

You can do that by performing the following steps:

1. Go to the query settings on the right pane of the Power Query UI and delete the **Filtered Rows** step.

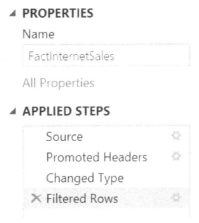

Figure 6.23 – Deleting the Filtered Rows step

2. Right-click on the ProductKey column and click on **Add as New Query**.

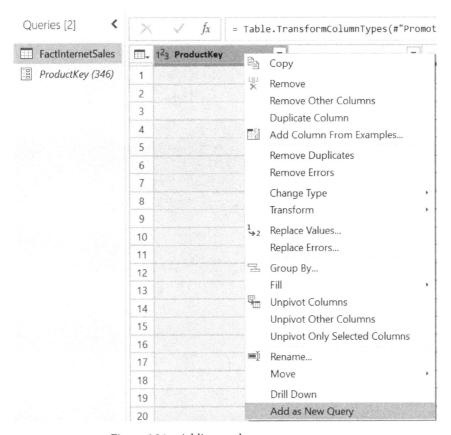

Figure 6.24 – Adding a column as a new query

3. A list will be generated with unique columns containing `ProductKey` values. Rename the query `ProductKeyList`.

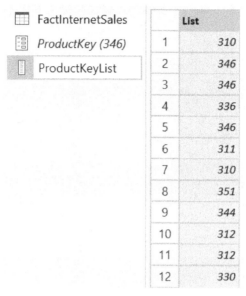

Figure 6.25 – Renaming a list

4. Right-click on **List** and click on **Remove Duplicates** in order to have unique values only.

Figure 6.26 – Remove Duplicates

5. Browse to the **Home** tab and click on **Manage Parameters**.

Figure 6.27 – Manage Parameters button

6. Edit the Product Key parameter by selecting **Query** from the drop-down section of the **Suggested Values** field.

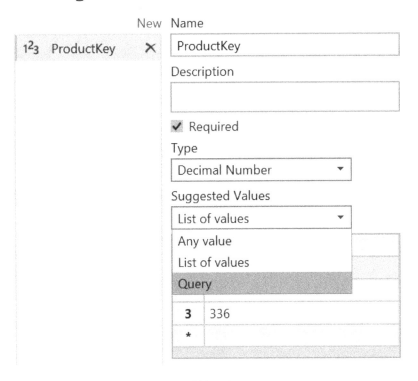

Figure 6.28 – Values from a query

7. Then, select **ProductKeyList** from the drop-down section of the **Query** field and click on **OK** to create the parameter.

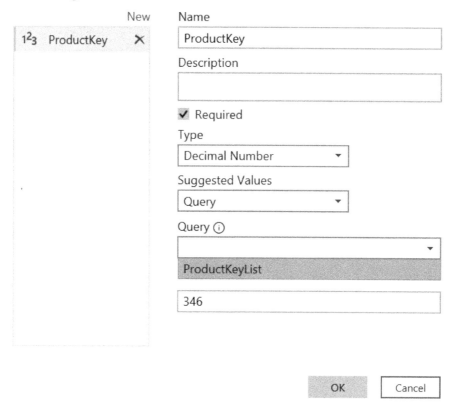

Figure 6.29 – Selecting ProductKeyList for Query

8. From the **Queries** pane, select the `FactInternetSales` query. Then, select the `ProductKey` column from the same query. Then, click on the drop-down icon on the right part of the `ProductKey` column and click on **Number Filters** and then **Equals…**, as shown in the following screenshot:

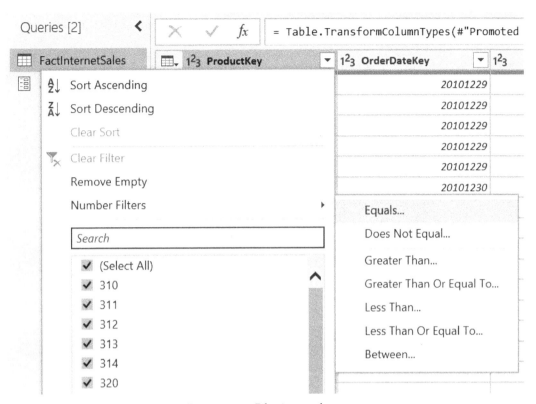

Figure 6.30 – Filtering a column

9. Click on the type of value you want to base your equality on and select **Parameter**.

Filter Rows

Apply one or more filter conditions to the rows in this table.

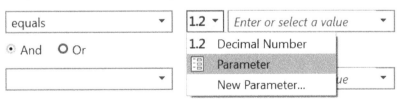

Figure 6.31 – Filter Rows window

10. **ProductKey** will be selected automatically since it's the only parameter in the current session. After this, click on **OK**.

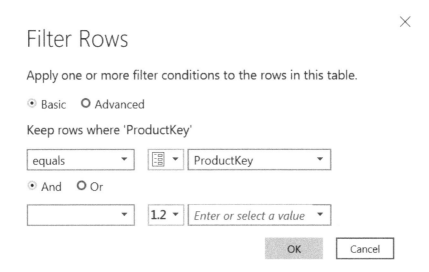

Figure 6.32 – Using a parameter as a filter

11. Browse to the **Home** tab, click on the **Manage Parameters** dropdown, and then click on **Edit Parameters**.

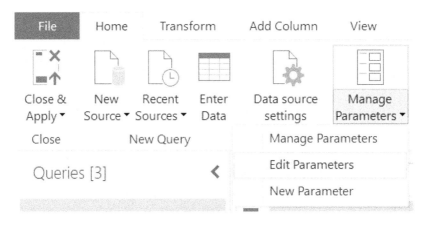

Figure 6.33 – Edit Parameters button

12. Click on the dropdown and observe how you can now choose among `ProductKey` values from the list we extracted from the `FactInternetSales` query.

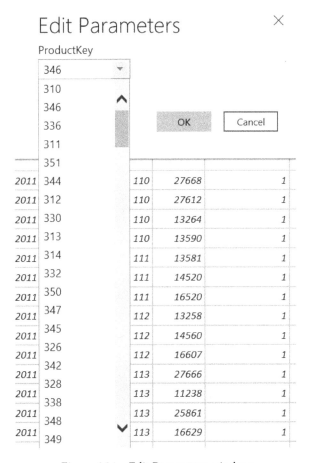

Figure 6.34 – Edit Parameters window

You can see how you can adapt parameters to different use cases. You can also create multiple parameters and apply combinations of filters to the same query.

Folding queries

You often connect to relational sources, and it is important to know how to leverage query folding in order to retrieve data from the sources with Power Query steps that act as a single query statement. **Query folding** helps us to push more steps toward the origin data source in order to reduce the number of steps processed by the Power Query engine.

In this recipe, you will see how to perform query folding and how to control it.

Getting ready

In this recipe, you need to connect to an **Azure SQL database** that you can recreate in your environment with the Adventureworks.bacpac file.

How to do it...

Once you open your Power BI Desktop application, you are ready to perform the following steps:

1. Click on **Get data** and then **More…** to access the **Get Data** window and see the complete list of connectors.

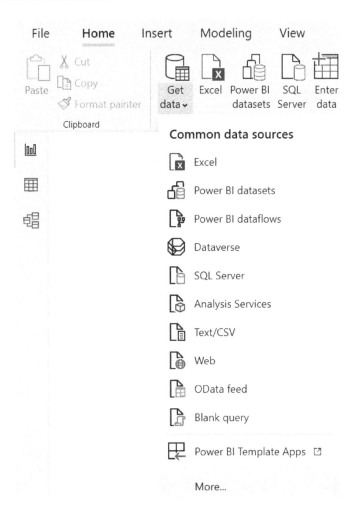

Figure 6.35 – Power BI connectors

2. Browse to the **Azure SQL database** connector, select it, and click on **Connect**.

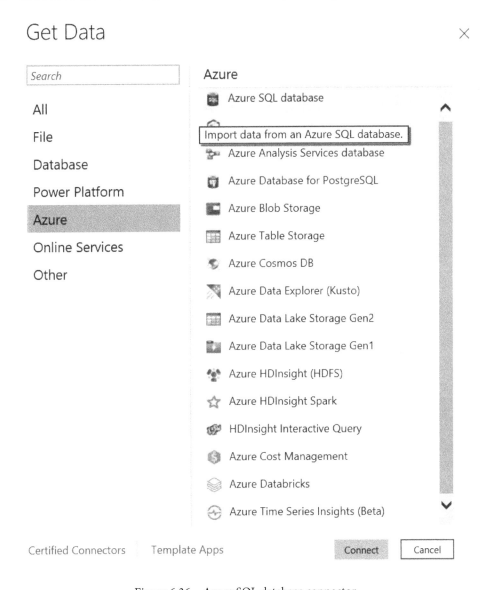

Figure 6.36 – Azure SQL database connector

3. Enter your server and database information, flag **Import** for **Data Connectivity mode**, and then click on **OK**.

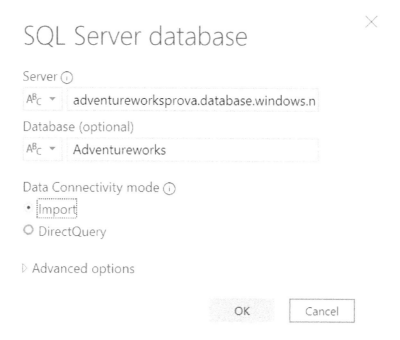

Figure 6.37 – SQL Server database information

4. Authenticate with your preferred authentication method. In this example, I'm using the **Microsoft account** authentication.

Figure 6.38 – SQL Server database authentication

5. Select the `FactInternetSales` table from the database and click on **Transform Data**.

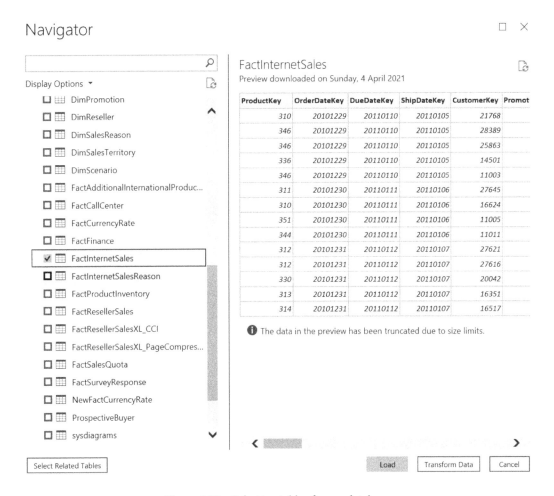

Figure 6.39 – Selecting tables from a database

6. Go to **APPLIED STEPS** and right-click on the **Navigation** step. You can see that you are able to select **View Native Query** and this means that query folding is now active:

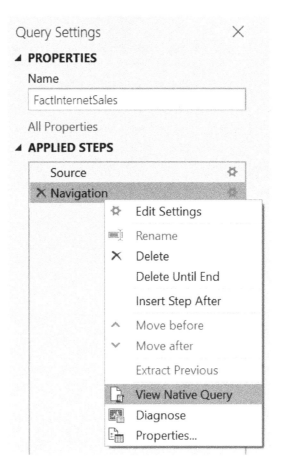

Figure 6.40 – View Native Query

7. Select the ProductKey column, click on the filter dropdown, select the first four values, as shown in the following screenshot, and click on **OK**:

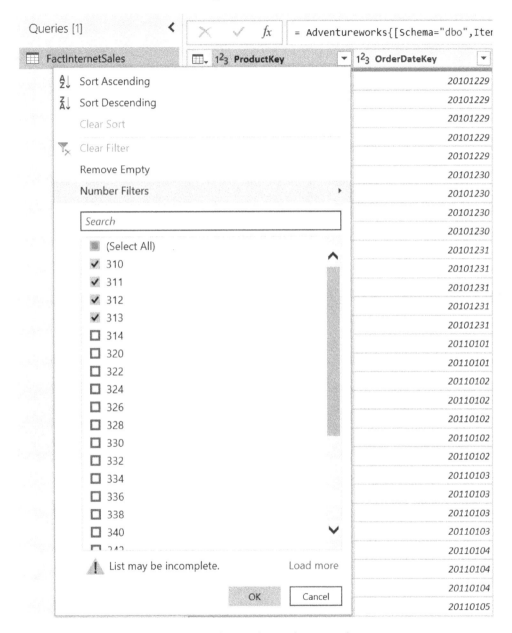

Figure 6.41 – Filtering the ProductKey column

8. Go to the **APPLIED STEPS** section in the right pane in the Power Query UI, right-click on the **Filtered Rows** step, as in *Step 6*, and click on **View Native Query**. The **Native Query** window will pop up, where you can see what the statement that is executed against the data source is:

Native Query   ☐ ✕

```
select [_].[ProductKey],
    [_].[OrderDateKey],
    [_].[DueDateKey],
    [_].[ShipDateKey],
    [_].[CustomerKey],
    [_].[PromotionKey],
    [_].[CurrencyKey],
    [_].[SalesTerritoryKey],
    [_].[SalesOrderNumber],
    [_].[SalesOrderLineNumber],
    [_].[RevisionNumber],
    [_].[OrderQuantity],
    [_].[UnitPrice],
    [_].[ExtendedAmount],
    [_].[UnitPriceDiscountPct],
    [_].[DiscountAmount],
    [_].[ProductStandardCost],
    [_].[TotalProductCost],
    [_].[SalesAmount],
    [_].[TaxAmt],
    [_].[Freight],
    [_].[CarrierTrackingNumber],
    [_].[CustomerPONumber],
    [_].[OrderDate],
    [_].[DueDate],
    [_].[ShipDate]
from [dbo].[FactInternetSales] as [_]
where ([_].[ProductKey] = 310 or [_].[ProductKey] = 311) or ([_].[ProductKey] = 312 or [_].[ProductKey] = 313)
```

OK

Figure 6.42 – Native query details

When you apply a filter, it is like you are applying a WHERE statement toward the database. You can also select columns, group data, merge queries with JOIN statements, pivot and unpivot, and achieve query folding. If you change the data type, you will see how query folding will be disabled. Go through the following example:

1. Go to the OrderDateKey column, browse to the **Transform** tab, and click on **Split Column** and **By Number of Characters**.

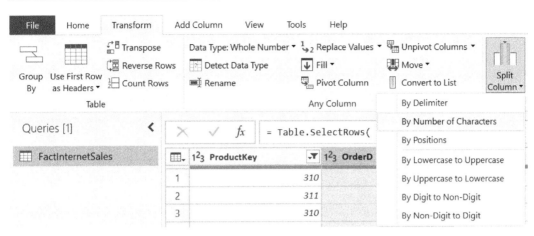

Figure 6.43 – Split Column button

2.   Enter 4 for **Number of characters**, flag **Once, as far left as possible**, and click on **OK**.

Split Column by Number of Characters

Specify the number of characters used to split the text column.

Number of characters

4

Split

◉ Once, as far left as possible

○ Once, as far right as possible

○ Repeatedly

▷ Advanced options

OK   Cancel

Figure 6.44 – Split Column by Number of Characters

3. Go to **APPLIED STEPS** and right-click on the **Navigation** step. You can see that you can't select **View Native Query** and this means that you can't leverage query folding, as shown in the following screenshot:

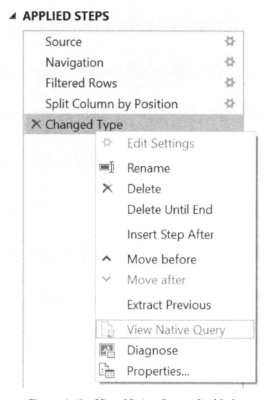

Figure 6.45 – View Native Query disabled

When you perform some changes on the data, when you add columns and enrich the content of your queries, you will probably lose the query folding feature. The best practice is to perform the steps when query folding is active at the beginning in order to send a single statement to the data source, and this will end up improving the overall performance, both in **Import** and **Direct Query** mode.

Leveraging incremental refresh and folding

When you load data from Power Query, you do not perform a one-time load, but usually you need to refresh data in order to load new data or to update existing data. When loading data incrementally, it is possible to leverage parameters and query folding in order to optimize and retrieve data quickly. In this recipe, we will see how to set up time parameters and incremental refresh.

Getting ready

In this recipe, you need to connect to an Azure SQL database that you can recreate in your environment with the `Adventureworks.bacpac` file.

How to do it...

Once you open your Power BI Desktop application, you are ready to perform the following steps:

1. Click on **Get data** and click on **More…** to access the **Get Data** window and to see the complete list of connectors.

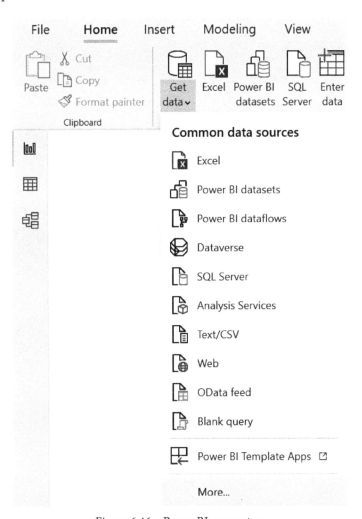

Figure 6.46 – Power BI connectors

2.   Browse to the **Azure SQL database** connector, select it, and click on **Connect**.

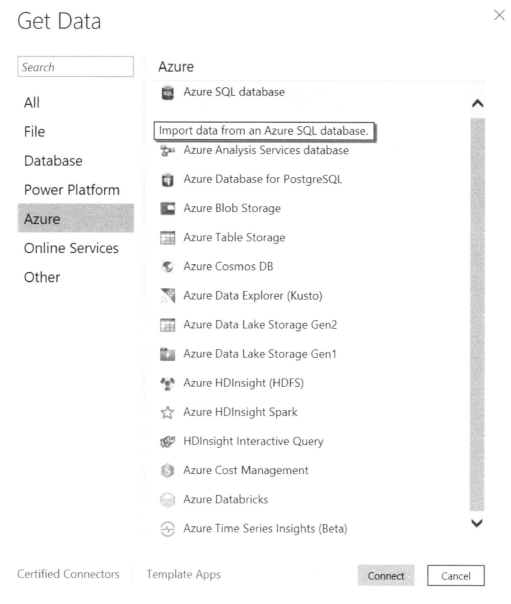

Figure 6.47 – Azure SQL database connector

3. Enter your server and database information, flag **Import** for **Data Connectivity mode**, and then click on **OK**.

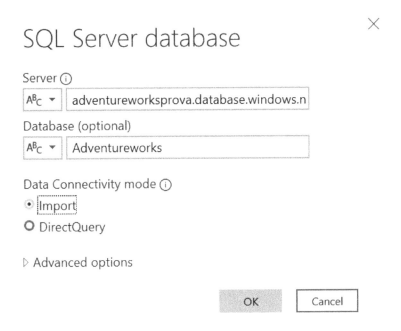

Figure 6.48 – SQL Server database information

4. Authenticate with your preferred authentication method. In this example, I'm using the **Microsoft account** authentication:

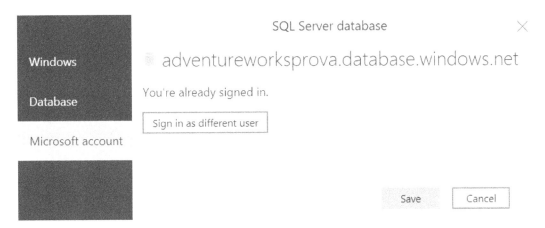

Figure 6.49 – SQL Server database authentication

5. Select the `FactInternetSales` table from the database and click on
    Transform Data.

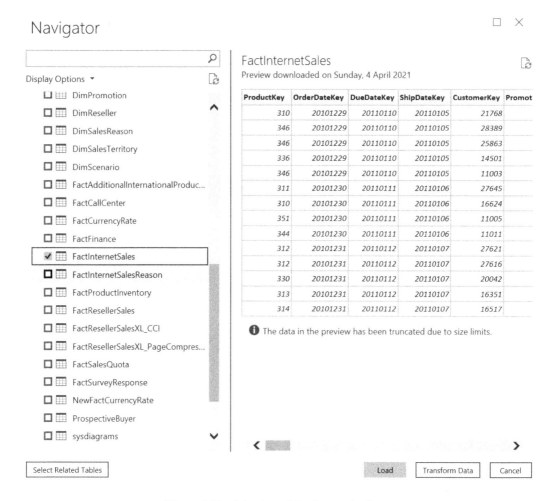

Figure 6.50 – Selecting tables from a database

6. Browse to the **Home** tab and click on the **Manage Parameters** button.

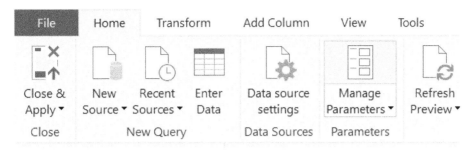

Figure 6.51 – Manage Parameters button

7. Create a `RangeStart` parameter with **Date/Time** for **Type** and **Any value** for **Suggested Values** and enter `29/12/2010 00:00:00` for **Current Value**, as shown in the following screenshot:

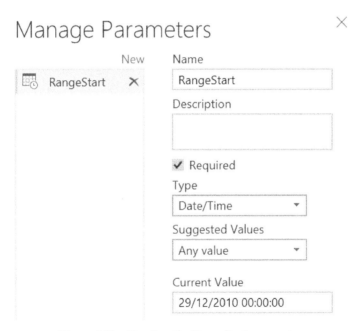

Figure 6.52 – Creating the RangeStart parameter

8. Create a second parameter, RangeEnd, with the same settings as the previous one, and then enter 05/01/2011 00:00:00 for **Current Value** and click on **OK**.

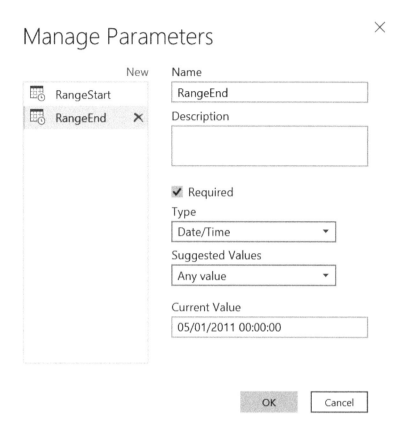

Figure 6.53 – Creating the RangeEnd parameter

9. Select the `FactInternetSales` query and select the `OrderDate` column. Click on the filter icon, then **Date/Time Filters**, and then **Custom Filter…**.

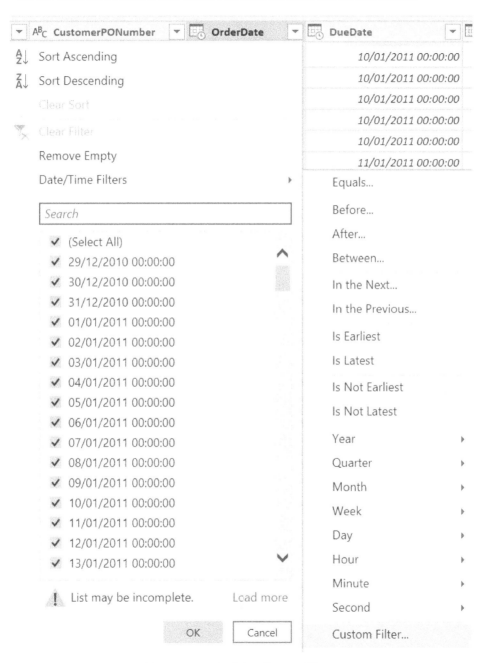

Figure 6.54 – Applying Custom Filter…

10. The **Filter Rows** window will pop up. Enter the two parameters as filter options, as shown in the following screenshot, and click on **OK** to apply the custom filters:

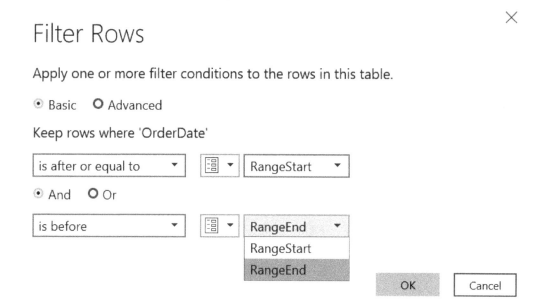

Figure 6.55 – Filter Rows window

In this way, we are defining a subset of data between a range defined by the parameters. The values of those parameters are not important at this time, because this acts just as a sample subset of data that will be managed by the incremental refresh rule that will be set later.

11. Click on **Close & Apply** to load all the queries within the model.

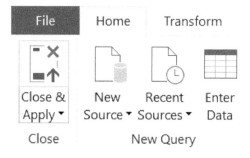

Figure 6.56 – Close & Apply

12. Go to the **Fields** section, right-click on the `FactInternetSales` table, and click on **Incremental refresh**.

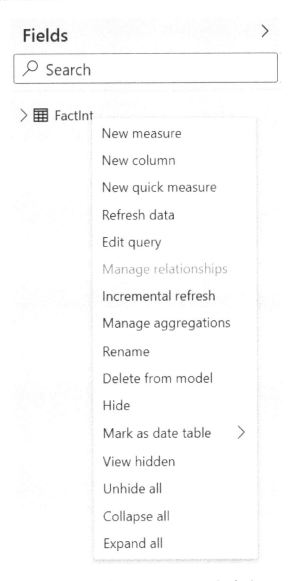

Figure 6.57 – Setting up incremental refresh

13. After having loaded the model with a sample subset of data defined by the two parameters, you can now set up a rule that will be used when you refresh the data from the Power BI service. You can select the table on which to apply the incremental refresh, enable the feature, and define which rows to store based on a historical period. In this case, we will store all data from the last 15 years. Then, you can define which time range to consider when refreshing data. In this case, we defined a 2-month refresh range, which means that RangeStart and RangeEnd, the two date parameters defined previously, will adapt to this rule and that subset of data will be updated and added. Once you set this up, leave unflagged the optional settings (**Detect data changes**, which allows you to update data that has changed, and **Only refresh complete months**), as shown in the following screenshot, and click on **Apply all**:

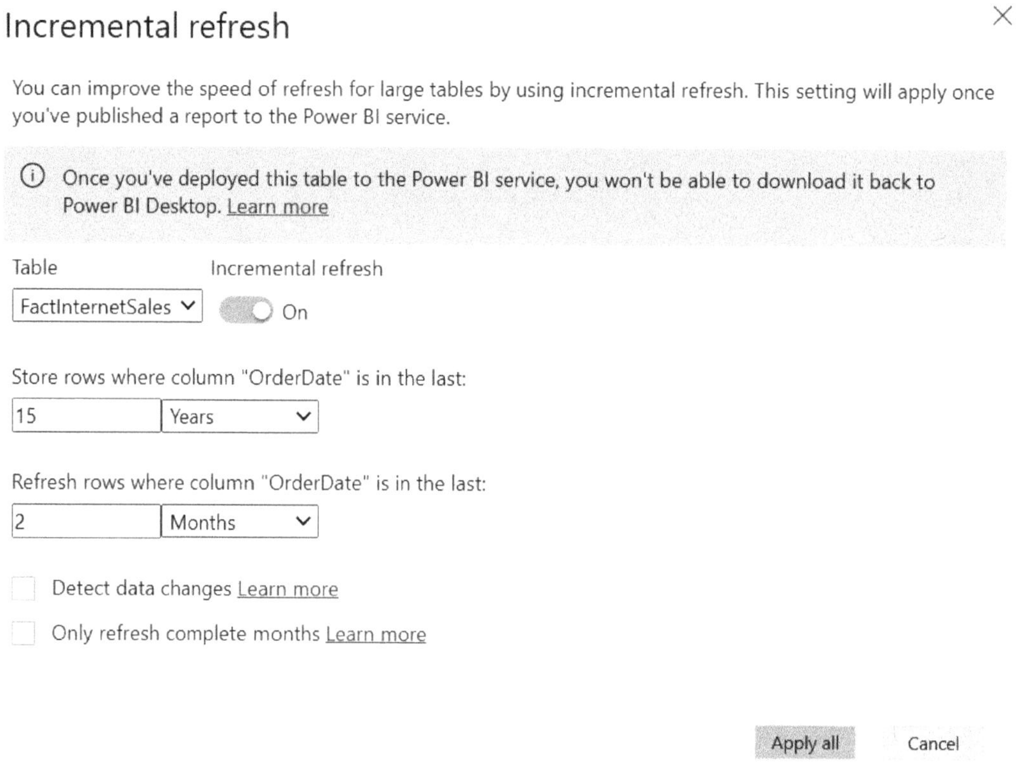

Incremental refresh   ✕

You can improve the speed of refresh for large tables by using incremental refresh. This setting will apply once you've published a report to the Power BI service.

ⓘ Once you've deployed this table to the Power BI service, you won't be able to download it back to Power BI Desktop. Learn more

Table   Incremental refresh

| FactInternetSales ∨ | ⬤ On

Store rows where column "OrderDate" is in the last:

| 15 | Years ∨ |

Refresh rows where column "OrderDate" is in the last:

| 2 | Months ∨ |

☐ Detect data changes Learn more

☐ Only refresh complete months Learn more

Apply all   Cancel

Figure 6.58 – Incremental refresh details

Once you publish the model on the Power BI service, you will be able to trigger the refresh from there and update data quickly. Query folding is key to achieve this functionality because when you send the refresh query to the source, you will be applying a filter that will be included in the statement sent to the source.

Disabling query load

Queries' loads can be heavy and sometimes refreshing some tables can impact negatively on performance. This is why it is important to know what data is concretely needed for end users. It is common that you will need some queries just for transformations in Power Query, but you won't need them in the final model. In this recipe, we will see how you can use some queries for enriching data needed for reporting and how you can disable the loading of this supporting table in order to reduce the impact on performance and refreshing.

Getting ready

For this recipe, you need to download the following files:

- The `FactInternetSales` CSV file
- The `DimTerritory` CSV file

In this example, we will refer to the `C:\Data` folder.

How to do it...

Once you open your Power BI Desktop application, you are ready to perform the following steps:

1. Click on **Get data** and select the **Text/CSV** connector.

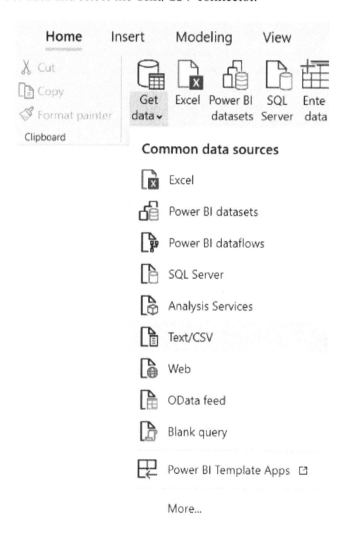

Figure 6.59 – Text/CSV connector

2. Browse to your local folder where you downloaded the `FactInternetSales.csv` file and open it. The following window with a preview of the data will pop up. Click on **Transform Data**.

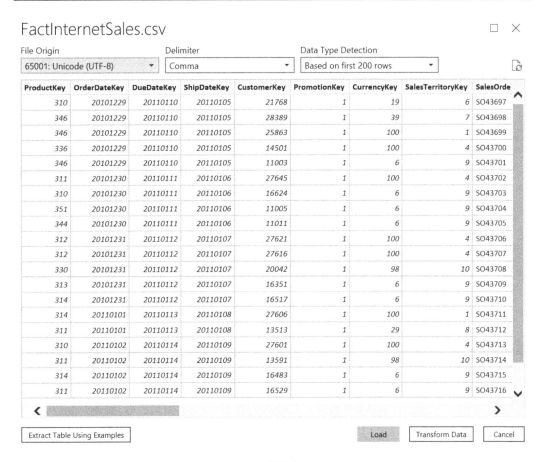

Figure 6.60 – CSV data preview

3. Repeat the previous two steps and load the `DimTerritory.csv` file.

4. Select **DimTerritory** in the **Queries** pane, browse to the **Home** tab, and click on the **Merge Queries** button.

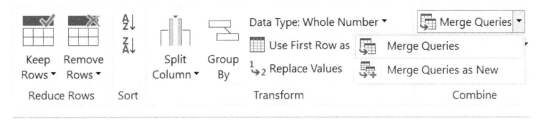

Figure 6.61 – Merge Queries button

5. The **Merge** window will pop up. Select `FactInternetSales` as the table to merge with `DimTerritory` and select the `SalesTerritoryKey` column from both tables. Select **Left Outer (all from first, matching from second)** and click on **OK**.

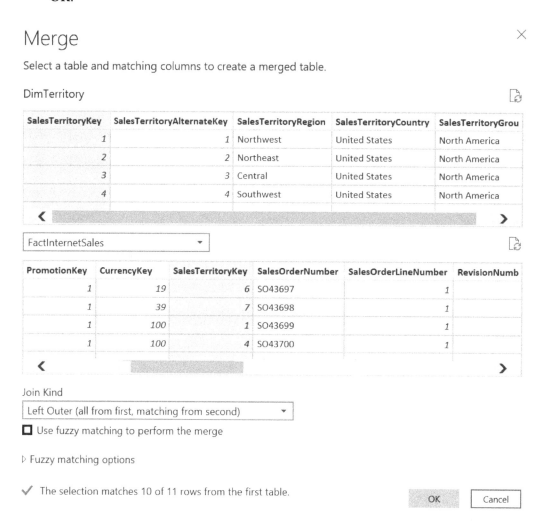

Figure 6.62 – Merge window

6. Click on the expand button on the right side of the `FactInternetSales` column. Flag **Aggregate**, select **Sum of TotalProductCost** and **Sum of SalesAmount**, and remove the flag from **Use original column name as prefix**, as shown in the following screenshot:

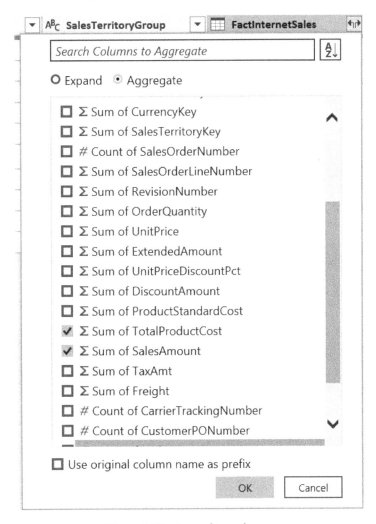

Figure 6.63 – Expanding columns

7. You should see the two newly added columns in the `DimTerritory` query.

ABC 123 Sum of TotalProductCost	ABC 123 Sum of SalesAmount
2130235,255	3649866,551
3629,7059	6532,4682
1649,9092	3000,8296
3346387,415	5718150,812
6906,4234	12238,8496
1147923,361	1977844,862
1557752,993	2644017,714
1706941,573	2894312,338
5375145,508	9061000,584
2001221,433	3391712,211
null	null

Figure 6.64 – Newly added columns

8. Rename the `DimTerritory` query to `Sales geography` since it will be the table that we will load in the data model.

Figure 6.65 – Renaming queries

9. Right-click on `FactInternetSales` in the **Queries** pane and observe how, if you load the queries with the current setting, you will load both of the queries because they have **Enable load** flagged.

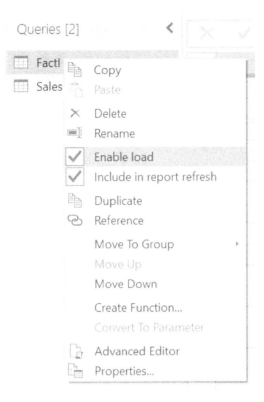

Figure 6.66 – Enable load button

10. If you click on **Enable load**, you will remove the flag and you will see that the name
of the query will turn into italics, which means that if you load the queries with
these settings, you will only be loading the `Sales geography` query and not
`FactInternetSales`, which in this case was used only to enrich the other query.

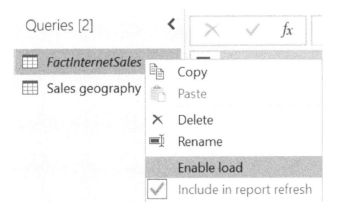

Figure 6.67 – Enable load button disabled

While using Power Query and performing data transformation steps, many queries will be used to enrich others and there is no concrete need to load them all. This will leave you with poor performance and high loading and refresh times and will increase the complexity of the data model.

By disabling the loading of some queries, you won't lose the transformations applied, such as the merge in this recipe, but you won't need to load a high-volume table.

7
Leveraging the M Language

Power Query is based on **M language**, which stands for **Power Query Formula Language**. Every time you perform a Power Query step, you are essentially writing M code. You can leverage the UI to transform your data without learning any M language at all but gaining an understanding of it could help you to customize even more Power Query transformations and perform quick corrections that are not possible with the UI only.

In this chapter, we will give an outline of M coding, explaining its differences from **Data Analysis Expression (DAX) language** (a familiar language to **Power BI** users), and you will see how to use M code on existing queries and how to create queries from scratch.

You will explore M coding examples through the following recipes:

- Using M syntax and the Advanced Editor
- Using M and DAX – differences
- Using M on existing queries
- Writing queries with M
- Creating tables in M
- Leveraging M – tips and tricks

Technical requirements

For this chapter, you will be using the following:

- **Power BI Desktop**: `https://www.microsoft.com/en-us/download/details.aspx?id=58494`

- Minimum requirements for installation:

  a) .NET Framework 4.6 (Gateway release August 2019 and earlier)

  b) .NET Framework 4.7.2 (Gateway release September 2019 and later)

  c) A 64-bit version of Windows 8 or a 64-bit version of Windows Server 2012 R2 with current TLS 1.2 and cipher suites

  d) 4 GB disk space for performance monitoring logs

You can find the data resources referred to in this chapter at the following link: `https://github.com/PacktPublishing/Power-Query-Cookbook/tree/main/Chapter07`.

Using M syntax and the Advanced Editor

Every step you perform in Power Query will translate into a line of M code. You usually realize what M code is after a while because you start using the features available from the UI at the beginning. Once you get more confident with Power Query steps, you become ready to explore the elements that lie behind them, learn how they work, and how you can create custom transformations by coding. In this recipe, we will see how to access M code and how steps are displayed in the **Advanced Editor**.

Getting ready

In this recipe, you need to download the `FactInternetSales.csv` file.

In this example, we will refer to the `C:\Data` folder.

How to do it...

Once you open your Power BI Desktop application, you are ready to perform the following steps:

1. Click on **Get Data** and select the **Text/CSV** connector:

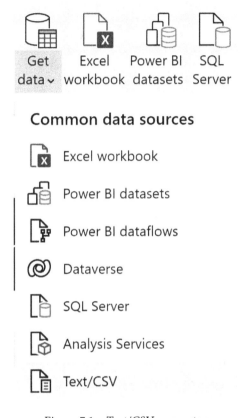

Figure 7.1 – Text/CSV connector

2. Browse to your local folder where you downloaded the `FactInternetSales.csv` file and open it. The following window, with a preview of the data, will pop up. Click on **Transform Data**:

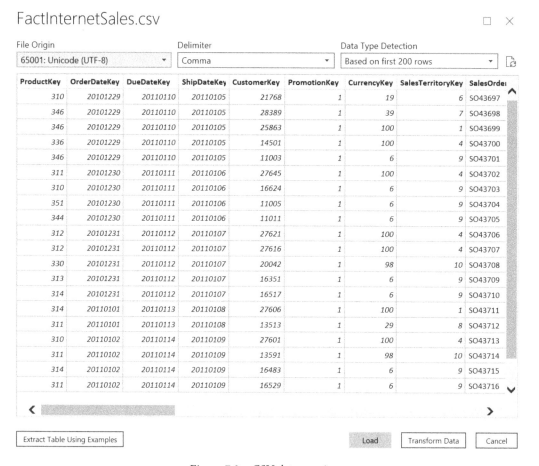

Figure 7.2 – CSV data preview

3. You will see the query displayed in the Power Query UI, and on the right, you will see some automatically generated steps, especially **Changed Type**. This detects data types, as you can see in the following screenshot:

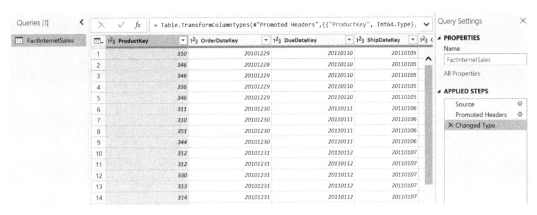

Figure 7.3 – Queries and Applied Steps

4. Imagine you want to modify the Changed Type step and apply a different data type to a column. In order to do this, you need to browse to the **Home** tab and click on **Advanced Editor**, as shown in the following screenshot:

Figure 7.4 – Advanced Editor button

5. The **Advanced Editor** window will pop up, and this will be the entry point in order to access the underlying M code. For each applied step listed in the UI, you can see that there is a line of code:

Figure 7.5 – Advanced Editor window

6. Each step can be identified with its name (`Source`, `Promoted Headers`, and `Changed Type`) and is separated from the previous one with a comma (`,`), as highlighted in the following screenshot:

FactInternetSales

```
let
    Source = Csv.Document(File.Contents("C:\Data\FactInternetSales.csv"),[Delimiter=",
        QuoteStyle=QuoteStyle.None]),
    #"Promoted Headers" = Table.PromoteHeaders(Source, [PromoteAllScalars=true]),
    #"Changed Type" = Table.TransformColumnTypes(#"Promoted Headers",{{"ProductKey", ]
        {"DueDateKey", Int64.Type}, {"ShipDateKey", Int64.Type}, {"CustomerKey", Int6
```

Figure 7.6 – Steps in the Advanced Editor

7. You can also change how to display your M code by clicking on the top right on **Display Options** and flag whether you want to **Display line numbers**, **Render whitespace**, **Display mini map**, and **Enable word wrap**. In this case, the second and last options were already flagged by default and in addition to this, we also flagged the first one to see the displayed code ordered in numbered lines:

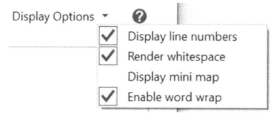

Figure 7.7 – Flagged Display line numbers

8. Now have a look at how the different steps are displayed. Every step contains a reference to the previous one. On line 3, you have the `Promoted Headers` step, and the first variable of the function is a reference to the previous step, `Source`:

```
1  let
2      Source = Csv.Document(File.Contents("C:\Data\FactInternetSales.csv"),[Delimiter=
           QuoteStyle=QuoteStyle.None]),
3      #"Promoted Headers" = Table.PromoteHeaders(Source, [PromoteAllScalars=true]),
4      #"Changed Type" = Table.TransformColumnTypes(#"Promoted Headers",{{"ProductKey",
           Int64.Type}, {"DueDateKey", Int64.Type}, {"ShipDateKey", Int64.Type}, {"Cust
           {"PromotionKey", Int64.Type}, {"CurrencyKey", Int64.Type}, {"SalesTerritoryK
           {"SalesOrderNumber", type text}, {"SalesOrderLineNumber", Int64.Type}, {"Rev
           {"OrderQuantity", Int64.Type}, {"UnitPrice", type number}, {"ExtendedAmount"
           {"UnitPriceDiscountPct", Int64.Type}, {"DiscountAmount", Int64.Type}, {"Prod
           {"TotalProductCost", type number}, {"SalesAmount", type number}, {"TaxAmt",
           number}, {"CarrierTrackingNumber", type text}, {"CustomerPONumber", type tex
           , {"DueDate", type datetime}, {"ShipDate", type datetime}})
5  in
6      #"Changed Type"
```

Figure 7.8 – Advanced Editor M code

9. Let's change the data type for the `ProductKey` column by selecting `Int64.Type`:

```
4   #"Changed Type" = Table.TransformColumnTypes(#"Promoted Headers",{{"ProductKey", Int64.Type}, {"OrderDateKey",
        Int64.Type}, {"DueDateKey", Int64.Type}, {"ShipDateKey", Int64.Type}, {"CustomerKey", Int64.Type},
        {"PromotionKey", Int64.Type}, {"CurrencyKey", Int64.Type}, {"SalesTerritoryKey", Int64.Type},
        {"SalesOrderNumber", type text}, {"SalesOrderLineNumber", Int64.Type}, {"RevisionNumber", Int64.Type},
        {"OrderQuantity", Int64.Type}, {"UnitPrice", type number}, {"ExtendedAmount", type number},
        {"UnitPriceDiscountPct", Int64.Type}, {"DiscountAmount", Int64.Type}, {"ProductStandardCost", type number},
        {"TotalProductCost", type number}, {"SalesAmount", type number}, {"TaxAmt", type number}, {"Freight", type
        number}, {"CarrierTrackingNumber", type text}, {"CustomerPONumber", type text}, {"OrderDate", type datetime}
        , {"DueDate", type datetime}, {"ShipDate", type datetime}})
5   in
6       #"Changed Type"
```

Figure 7.9 – Advanced Editor change type

10. Now replace `Int64.Type` with `type text` and click on **Done**:

```
4   #"Changed Type" = Table.TransformColumnTypes(#"Promoted Headers",{{"ProductKey", type text}, {"OrderDateKey",
        Int64.Type}, {"DueDateKey", Int64.Type}, {"ShipDateKey", Int64.Type}, {"CustomerKey", Int64.Type},
        {"PromotionKey", Int64.Type}, {"CurrencyKey", Int64.Type}, {"SalesTerritoryKey", Int64.Type},
        {"SalesOrderNumber", type text}, {"SalesOrderLineNumber", Int64.Type}, {"RevisionNumber", Int64.Type},
        {"OrderQuantity", Int64.Type}, {"UnitPrice", type number}, {"ExtendedAmount", type number},
        {"UnitPriceDiscountPct", Int64.Type}, {"DiscountAmount", Int64.Type}, {"ProductStandardCost", type number},
        {"TotalProductCost", type number}, {"SalesAmount", type number}, {"TaxAmt", type number}, {"Freight", type
        number}, {"CarrierTrackingNumber", type text}, {"CustomerPONumber", type text}, {"OrderDate", type datetime}
        , {"DueDate", type datetime}, {"ShipDate", type datetime}})
5   in
6       #"Changed Type"
```

✓ No syntax errors have been detected.

[Done] [Cancel]

Figure 7.10 – Advanced Editor change type to text

You can see how the `ProductKey` column changed type, but no additional step was added because we edited the only step that was there in the **APPLIED STEPS** list:

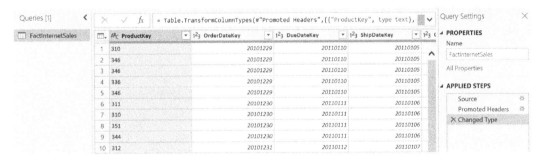

Figure 7.11 – ProductKey column with updated type

For this recipe, we connected to a CSV file, but imagine you need to connect to a database with the same data and perform the same transformations, but you do not want to create a query from scratch and perform all the steps again. In this case, you can leverage the Advanced Editor and change the source through the M code, following the *Authentication to data sources* recipe you will find in *Chapter 1, Getting Started with Power Query*.

Using M and DAX – differences

Generally, when you are presented with Power BI as a business intelligence tool and start exploring Power Query in it, you will use two code languages that have similar functionalities – DAX and M code – and you will probably get confused by these similarities. In this recipe, we will discuss the main differences between these two languages and when it is better to use one instead of the other. To illustrate the main differences between them, we will create an additional column with both M code and DAX.

Getting ready

In this recipe, you need to download the `FactInternetSales.csv` file.

In this example, we will refer to the `C:\Data` folder.

How to do it...

Once you open your Power BI Desktop application, you are ready to perform the following steps:

1. Click on **Get Data** and select the **Text/CSV** connector.

2. Browse to your local folder where you downloaded the `FactInternetSales.csv` file and open it. The following window with a preview of the data will pop up. Click on **Transform Data**.

3. Browse to the **Add Column** tab and click on **Custom Column**:

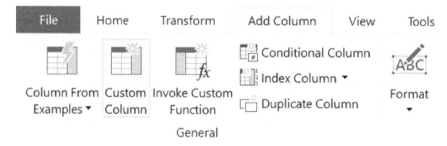

Figure 7.12 – Add custom column

4. Create a custom column named `Gross Margin` (intended here as the subtraction of `TotalProductCost` from `SalesAmount`) and enter the following formula:

    ```
    = [SalesAmount] - [TotalProductCost]
    ```

You can see in the following screenshot how the subtraction is executed. Now click on **OK**:

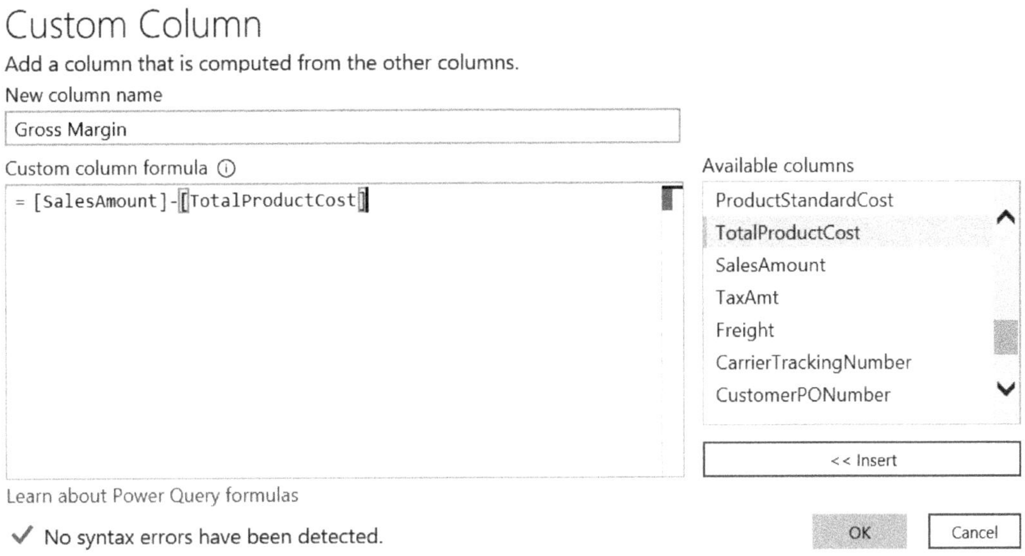

Custom Column

Add a column that is computed from the other columns.

New column name

Gross Margin

Custom column formula ⓘ

= [SalesAmount]-[TotalProductCost]

Available columns

ProductStandardCost
TotalProductCost
SalesAmount
TaxAmt
Freight
CarrierTrackingNumber
CustomerPONumber

<< Insert

Learn about Power Query formulas

✓ No syntax errors have been detected.

OK   Cancel

Figure 7.13 – Custom column definition

5.   Change the data type of this newly created column to **Decimal Number**:

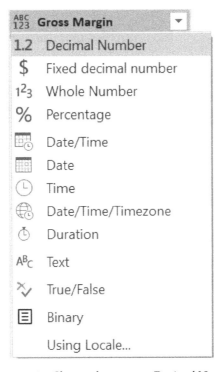

Figure 7.14 – Change data type to Decimal Number

6. You can see a newly created column as in the following screenshot:

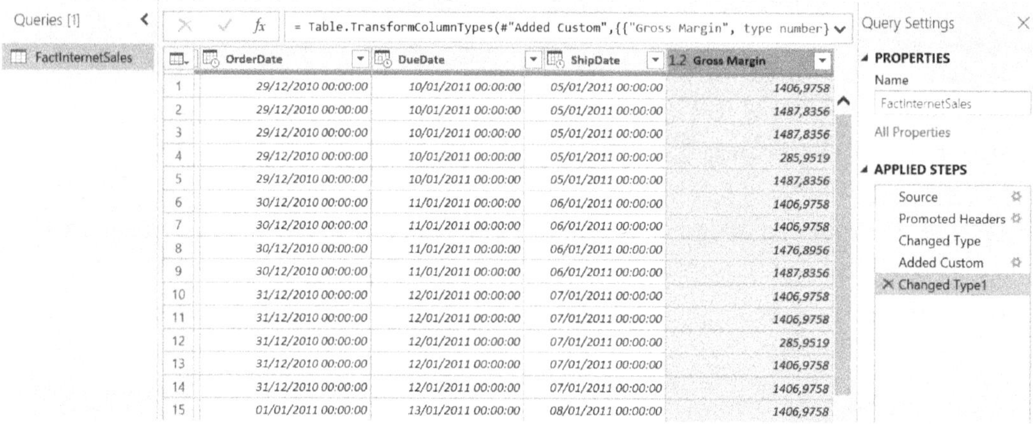

Figure 7.15 – Gross Margin calculated column

7. Open the **Advanced Editor** and observe the newly added line of code with the previous transformations:

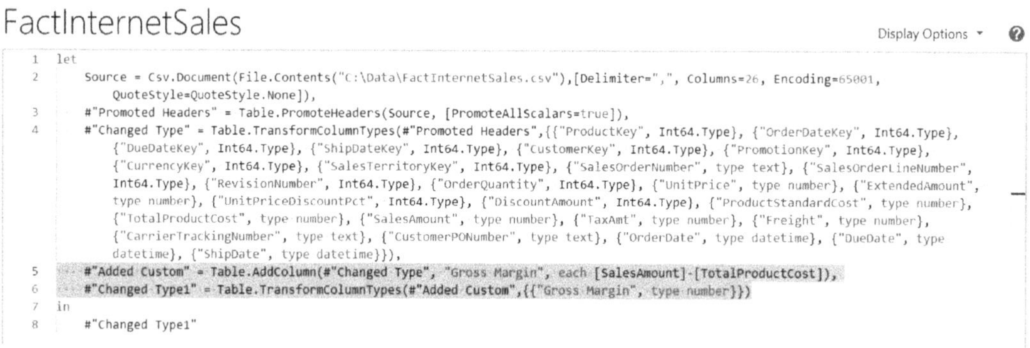

FactInternetSales

Display Options ▾ ❓

```
1  let
2      Source = Csv.Document(File.Contents("C:\Data\FactInternetSales.csv"),[Delimiter=",", Columns=26, Encoding=65001,
          QuoteStyle=QuoteStyle.None]),
3      #"Promoted Headers" = Table.PromoteHeaders(Source, [PromoteAllScalars=true]),
4      #"Changed Type" = Table.TransformColumnTypes(#"Promoted Headers",{{"ProductKey", Int64.Type}, {"OrderDateKey", Int64.Type},
          {"DueDateKey", Int64.Type}, {"ShipDateKey", Int64.Type}, {"CustomerKey", Int64.Type}, {"PromotionKey", Int64.Type},
          {"CurrencyKey", Int64.Type}, {"SalesTerritoryKey", Int64.Type}, {"SalesOrderNumber", type text}, {"SalesOrderLineNumber",
          Int64.Type}, {"RevisionNumber", Int64.Type}, {"OrderQuantity", Int64.Type}, {"UnitPrice", type number}, {"ExtendedAmount",
          type number}, {"UnitPriceDiscountPct", Int64.Type}, {"DiscountAmount", Int64.Type}, {"ProductStandardCost", type number},
          {"TotalProductCost", type number}, {"SalesAmount", type number}, {"TaxAmt", type number}, {"Freight", type number},
          {"CarrierTrackingNumber", type text}, {"CustomerPONumber", type text}, {"OrderDate", type datetime}, {"DueDate", type
          datetime}, {"ShipDate", type datetime}}),
5      #"Added Custom" = Table.AddColumn(#"Changed Type", "Gross Margin", each [SalesAmount]-[TotalProductCost]),
6      #"Changed Type1" = Table.TransformColumnTypes(#"Added Custom",{{"Gross Margin", type number}})
7  in
8      #"Changed Type1"
```

Figure 7.16 – Advanced Editor steps

8. Browse to the **Home** tab and click on **Close & Apply** to load the data and access the Power BI interface:

Figure 7.17 – Close & Apply button

Keep in mind that at this phase we have not imported or loaded anything in the data model yet. This custom column was created, its step was mapped in the Power Query code, and it was created as an output that comes from two inputs (SalesAmount and TotalProductCost) from the same table.

Let's see how we can create the same Gross Margin column with DAX by performing the following steps:

1. Browse to the **Modeling** tab and click on **New column**:

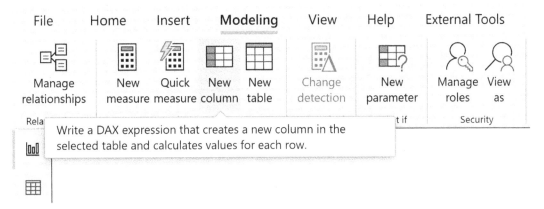

Figure 7.18 – Power BI view

The DAX bar will appear. Enter the following expression in the DAX bar to create a calculated column named Gross Margin 2:

```
Gross Margin 2 = FactInternetSales[SalesAmount]-FactInter
netSales[TotalProductCost]
```

2. Click on the **Data** tab on the left section of the Power Query UI, the second one displayed in the following screenshot:

Figure 7.19 – Data tab

Browse to the last two columns, Gross Margin and Gross Margin 2, and check that you can see the same values:

Gross Margin ▼	Gross Margin 2 ▼
3,1237	3,1237
3,1237	3,1237
3,1237	3,1237
3,1237	3,1237
3,1237	3,1237
3,1237	3,1237
3,1237	3,1237
3,1237	3,1237
3,1237	3,1237
3,1237	3,1237
3,1237	3,1237
3,1237	3,1237
3,1237	3,1237
3,1237	3,1237
3,1237	3,1237
3,1237	3,1237

Figure 7.20 – Gross Margin 2 as DAX calculated column

Gross Margin was calculated in the Power Query interface and Gross Margin 2 with DAX language inside Power BI. So, what is the difference between them if they are displaying the same values? The differences can be summed up as follows:

- **Power Query**: The first column was created in Power Query as a query-time transformation aimed to shape the data while extracting it from the data source. The steps are mapped, and it is considered a programming language with **M IntelliSense**, a code-completion aid that helps users with formula completion suggestions. You can perform steps through UI features and perform data preparation intuitively.

- **DAX**: The second column was created in Power BI after data was loaded as an in-memory transformation to analyze data after having extracted it. With **DAX**, which stands for **Data Analysis Expression**, it is possible to create formulas that also refer to other tables within the same data model. There is no trace of the steps performed (as in Power Query) because it is meant to be used to quickly address business challenges on top of an in-memory engine. It is a formula language, not a programming language.

Using M on existing queries

In the previous recipes, you saw how to change and edit steps by simply opening the Advanced Editor and modifying data types without adding additional steps or changing data sources, without the need of performing steps from scratch. In this recipe, we will see additional possibilities of how to use M code on existing queries and with few steps.

Getting ready

In this recipe, you need to connect to the **Azure SQL Database** that you can recreate in your environment with the Adventureworks.bacpac file.

How to do it...

Once you open your Power BI Desktop application, you are ready to perform the following steps:

1. Click on **Get Data** and click on **More…**:

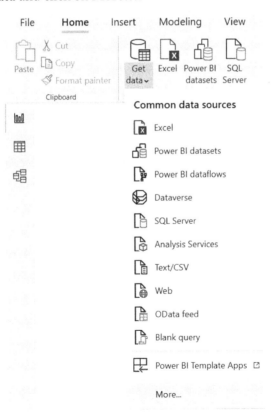

Figure 7.21 – Power BI connectors

2. Browse to the **Azure SQL database** connector, select it, and click on **Connect**:

Figure 7.22 – Azure SQL database connector

3. Enter your server and database information, flag **Import** as **Data Connectivity mode**, and then click on **OK**:

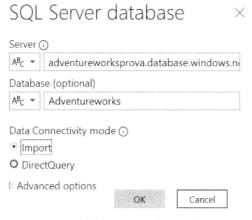

Figure 7.23 – SQL Server database information

4. Authenticate with your preferred authentication method. In this example, we're using the **Microsoft account** authentication:

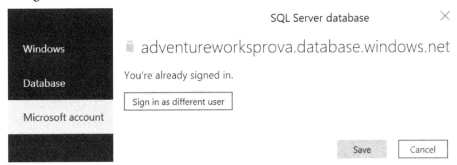

Figure 7.24 – SQL Server database authentication

5. Select the `FactInternetSales` table from the database and click on **Transform Data**:

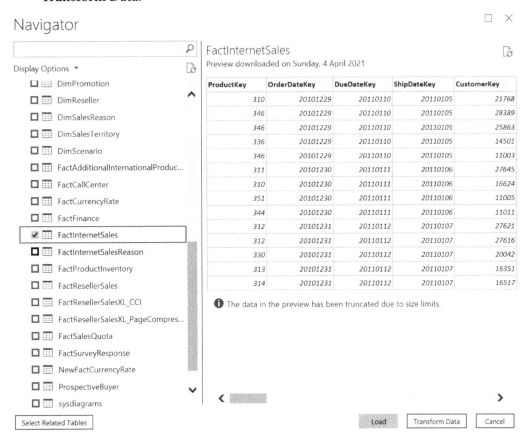

Figure 7.25 – Select tables from database

6. Click on **Choose Columns** and flag the following columns: **ProductKey**, **SalesTerritoryKey**, **TotalProductKey**, **SalesAmount**, and **OrderDate** and click on **OK**:

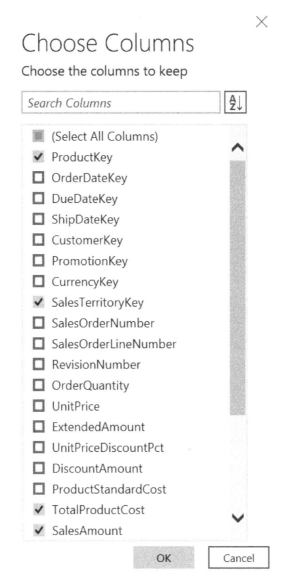

Figure 7.26 – Choose Columns window

7. You can see that for each step, you can see its M code in the formula bar above the data in the UI:

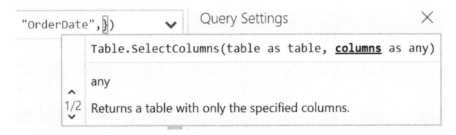

Figure 7.27 – Formula bar expression

8. Let's say that you want to add another column you have missed in the previous step. You could do that in the UI, but also directly with code. Click with your cursor after "Order Date" and add a comma (,), and observe how a window will pop up explaining how to interpret the M formula:

```
"OrderDate", )

Table.SelectColumns(table as table, columns as any)

any

1/2  Returns a table with only the specified columns.
```

Query Settings

Figure 7.28 – IntelliSense example

9. After the comma, add the value "DueDate", press *Enter* on your keyboard and see how the column DueDate appears in the query:

```
sTerritoryKey", "TotalProductCost", "SalesAmount", "OrderDate","DueDate"})
```

1.2 SalesAmount	OrderDate	DueDate
3578,27	29/12/2010 00:00:00	10/01/2011 00:00:00
3399,99	29/12/2010 00:00:00	10/01/2011 00:00:00
3399,99	29/12/2010 00:00:00	10/01/2011 00:00:00
699,0982	29/12/2010 00:00:00	10/01/2011 00:00:00
3399,99	29/12/2010 00:00:00	10/01/2011 00:00:00

Figure 7.29 – DueDate column added

10. Now we want to reorder columns. We could drag and drop columns with our mouse, use the Advanced Editor only, or use a combination of both. Select the DueDate column and drag it before OrderDate:

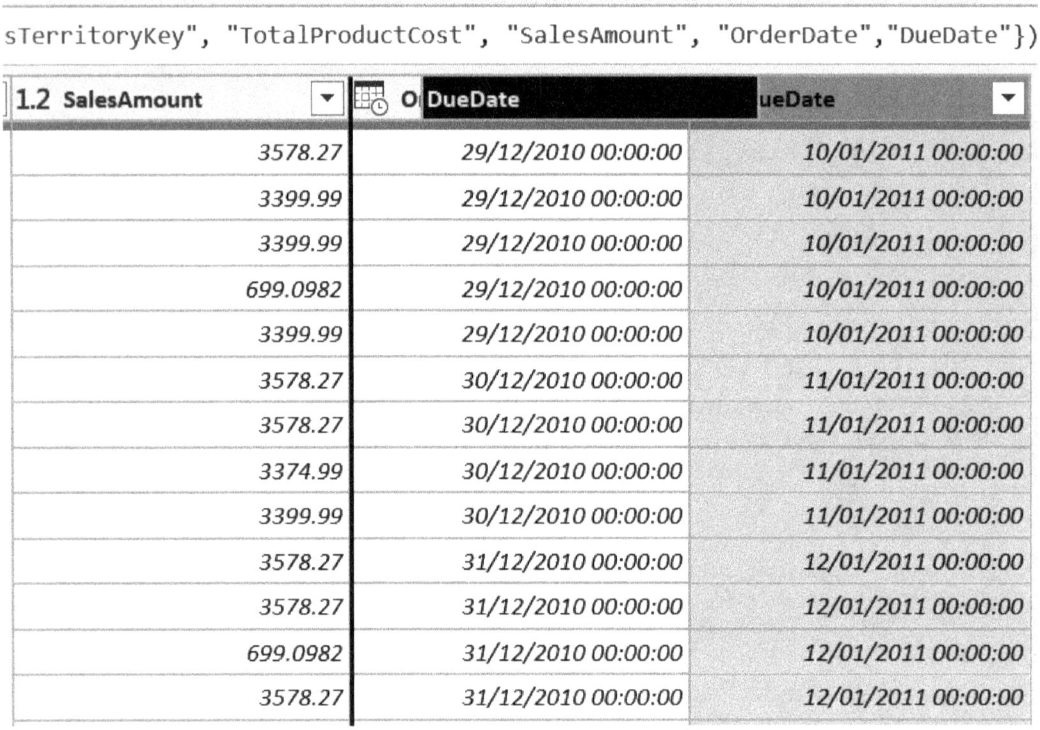

sTerritoryKey", "TotalProductCost", "SalesAmount", "OrderDate","DueDate"})

1.2 SalesAmount	O DueDate	ueDate
3578.27	29/12/2010 00:00:00	10/01/2011 00:00:00
3399.99	29/12/2010 00:00:00	10/01/2011 00:00:00
3399.99	29/12/2010 00:00:00	10/01/2011 00:00:00
699.0982	29/12/2010 00:00:00	10/01/2011 00:00:00
3399.99	29/12/2010 00:00:00	10/01/2011 00:00:00
3578.27	30/12/2010 00:00:00	11/01/2011 00:00:00
3578.27	30/12/2010 00:00:00	11/01/2011 00:00:00
3374.99	30/12/2010 00:00:00	11/01/2011 00:00:00
3399.99	30/12/2010 00:00:00	11/01/2011 00:00:00
3578.27	31/12/2010 00:00:00	12/01/2011 00:00:00
3578.27	31/12/2010 00:00:00	12/01/2011 00:00:00
699.0982	31/12/2010 00:00:00	12/01/2011 00:00:00
3578.27	31/12/2010 00:00:00	12/01/2011 00:00:00

Figure 7.30 – Drag DueDate column

11. You will again see how a step was created in the formula bar. You can now edit this step to define your own order. In this case, we will define the following order:

```
= Table.ReorderColumns(#"Removed Other
Columns",{"SalesTerritoryKey", "ProductKey",
"SalesAmount","TotalProductCost",  "DueDate",
"OrderDate"})
```

Let's now create a `flag` column that will define whether to apply a discount to a transaction or not. The values of this new column will be `Apply discount` or `Don't apply discount` and one or the other value will refer to each row depending on the value obtained by subtracting `TotalProductCost` from `SalesAmount`. If the net sales are higher than `1000`, then it will be possible to apply the discount and if not, it won't be applied. Click on **Custom Column** and create a new column, naming it `Discount status`:

Custom Column   ✕

Add a column that is computed from the other columns.

New column name

Discount status

Custom column formula ⓘ   Available columns

```
= if ([SalesAmount]-[TotalProductCost]) > 1000 then "Apply
  discount" else "Don't apply discount"
```

SalesTerritoryKey

ProductKey

SalesAmount

TotalProductCost

DueDate

OrderDate

Subtraction

<< Insert

Learn about Power Query formulas

✓ No syntax errors have been detected.   OK   Cancel

Figure 7.31 – Custom Column Discount status

Enter the following code and then click on **OK** to create the new custom column:

```
if ([SalesAmount]-[TotalProductCost]) > 1000 then "Apply
discount" else "Don't apply discount"
```

12. You can see the whole line of code on the formula bar and the new column added:

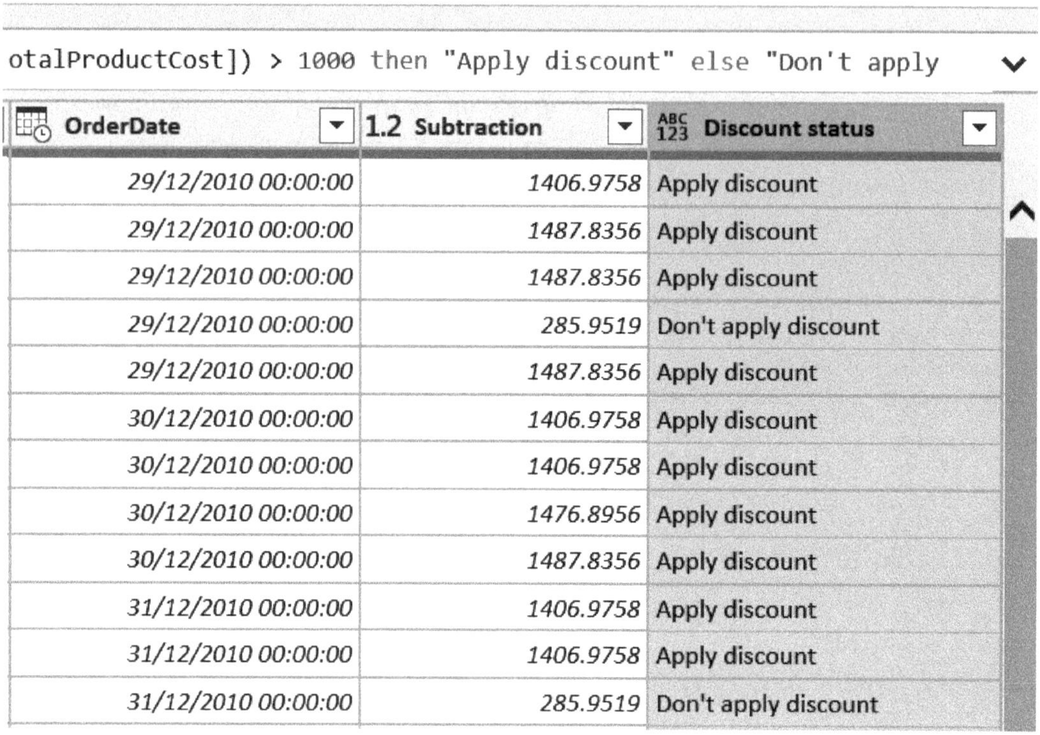

```
otalProductCost]) > 1000 then "Apply discount" else "Don't apply    ⌄
```

🕐 OrderDate	▼	1.2 Subtraction	▼	ABC 123 Discount status	▼
29/12/2010 00:00:00		1406.9758		Apply discount	
29/12/2010 00:00:00		1487.8356		Apply discount	
29/12/2010 00:00:00		1487.8356		Apply discount	
29/12/2010 00:00:00		285.9519		Don't apply discount	
29/12/2010 00:00:00		1487.8356		Apply discount	
30/12/2010 00:00:00		1406.9758		Apply discount	
30/12/2010 00:00:00		1406.9758		Apply discount	
30/12/2010 00:00:00		1476.8956		Apply discount	
30/12/2010 00:00:00		1487.8356		Apply discount	
31/12/2010 00:00:00		1406.9758		Apply discount	
31/12/2010 00:00:00		1406.9758		Apply discount	
31/12/2010 00:00:00		285.9519		Don't apply discount	

Figure 7.32 – Discount status column added

You can see how you can edit and enrich your existing data by leveraging the M code in Power Query. Editing existing queries and interacting with the Advanced Editor will help you to get confident with how this language works.

Writing queries with M

Once you become more confident with modifying existing steps that were created using the UI, you can further explore M language by creating elements such as values, lists, or tables without connecting to data sources, but directly coding from the Advanced Editor. In this recipe, we will see how M language works when writing queries from scratch. You will perform numerical operations, define variables, and create lists and simple tables.

Getting ready

In this recipe, you need to download the FactInternetSales.csv file.

In this example, we will refer to the C:\Data folder.

How to do it...

Once you open your Power BI Desktop application, you are ready to perform the following steps:

1. Click on **Get Data** and select the **Blank query** connector:

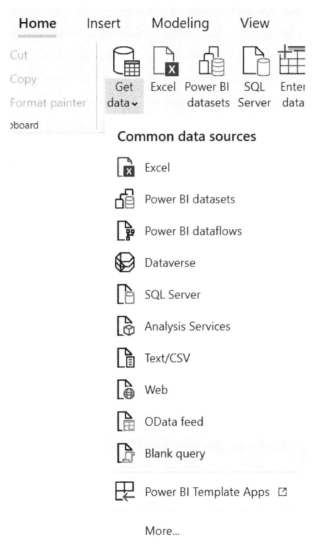

Figure 7.33 – Text/CSV connector

2. The Power Query UI will pop up and you will see an empty query with its default name as **Query1**:

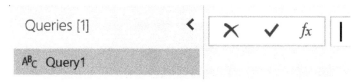

Figure 7.34 – Blank query

3. Browse to the **Home** tab and click on **Advanced Editor**.

4. You will see the code of the blank query and from here we can start experimenting with the potential of M code:

Query1

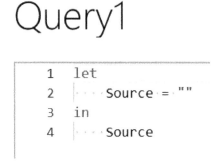

Figure 7.35 – Blank query in Advanced Editor

5. Let's introduce a subset of the code that will define the final output of the query that is established by the `in` expression. You can define several steps, and each step has a variable name to which an expression is assigned. Let's see an example of defining two numeric variables and an output that is their sum. Write the following example code:

```
let
    a= 1,
    b=2,
    sum= a+b
in
    sum
```

Click on **Done** and check the following screenshot:

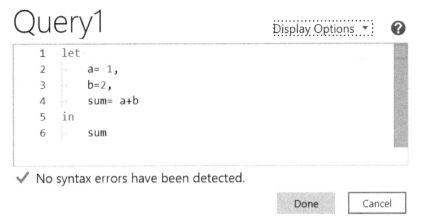

Figure 7.36 – Custom M expression

6. Observe the output in the Power Query UI to see the output value displayed at the center and the three steps in the **APPLIED STEPS** list explicated as in any other Power Query example seen earlier:

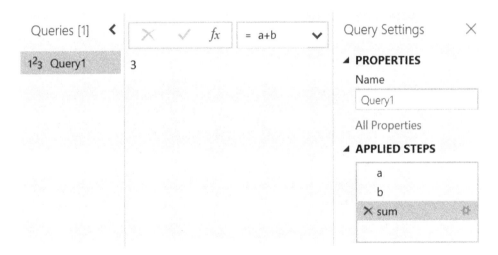

Figure 7.37 – Custom steps in the UI

7. If you click on the other steps, **a** or **b**, you will see their values displayed at the
    center, as in the following screenshot:

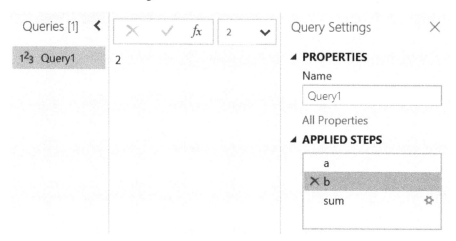

Figure 7.38 – Explore steps

What if you want to perform this sum with some values coming from other tables?
Have a look at the following steps to see how to do that:

1. Click on **New Source** and select **Text/CSV**:

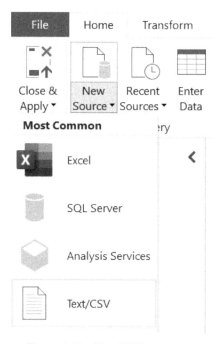

Figure 7.39 – Text/CSV connector

2. Browse to your local folder where you downloaded the `FactInternetSales.csv` file and open it. Click on **Choose Columns**, flag **SalesAmount**, and click on **OK**:

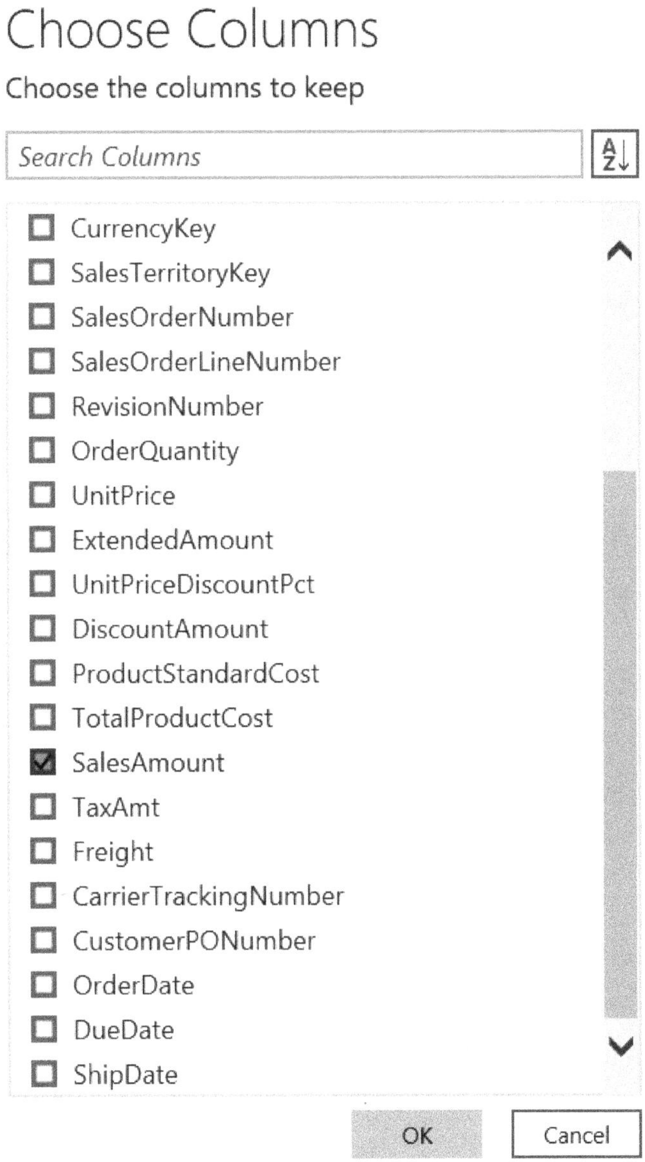

Choose Columns

Choose the columns to keep

Search Columns	A Z↓

- ☐ CurrencyKey
- ☐ SalesTerritoryKey
- ☐ SalesOrderNumber
- ☐ SalesOrderLineNumber
- ☐ RevisionNumber
- ☐ OrderQuantity
- ☐ UnitPrice
- ☐ ExtendedAmount
- ☐ UnitPriceDiscountPct
- ☐ DiscountAmount
- ☐ ProductStandardCost
- ☐ TotalProductCost
- ☑ SalesAmount
- ☐ TaxAmt
- ☐ Freight
- ☐ CarrierTrackingNumber
- ☐ CustomerPONumber
- ☐ OrderDate
- ☐ DueDate
- ☐ ShipDate

OK   Cancel

Figure 7.40 – Choose Columns window

3. Browse to the **Transform** tab, click on **Statistics** and then **Sum**, as shown in the following screenshot:

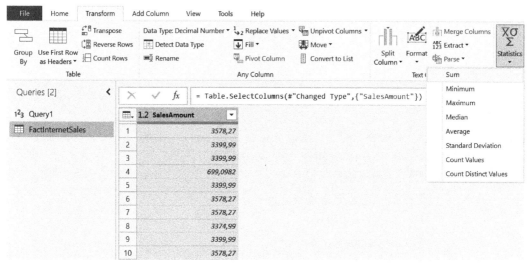

Figure 7.41 – Sum column values

4. A sum of all values in the column will be performed. Rename `FactInternetSales` to `SalesAmount`:

Figure 7.42 – SalesAmount value

5. Repeat all the previous steps (from 1 to 4), flagging **TotalProductCost** at *Step 2* and renaming the query to `TotalProductSales`:

Figure 7.43 – TotalProductCost value

6. Now open the **Advanced Editor** of **Query1** and write the following:

```
let
    Discount= 0.2,
    NetSales= SalesAmount - TotalProductSales,
    NetSales_discounted= NetSales - NetSales * Discount
in
    NetSales_discounted
```

With these few lines, we will retrieve values calculated in other queries (`SalesAmount` and `TotalProductSales`) and use them to calculate a `NetSales` expression. We will then apply an operation to calculate a discounted value that in this case is a fixed value (`0.2`). Once you apply these steps, click on **Done**:

Figure 7.44 – Advanced Editor custom query

7. Rename the `Query1` expression to `NetSales_discounted` and observe the calculated value:

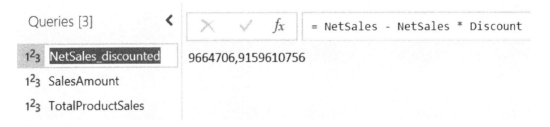

Figure 7.45 – NetSales_discounted final output

You can write both static and dynamic logic with M code as in the preceding examples. You calculated two values and then you retrieved those in a custom query.

You can also create lists with few lines of code. Follow these steps to see how to do it:

1. Click on **New Source** and then on **Blank Query**:

Figure 7.46 – Blank Query connector

2. Open the **Advanced Editor**, write the following code and click on **Done:**

```
let
    List= {1,2,3}
in
    List
```

3. You can see how a list was created with the values defined in the previous step:

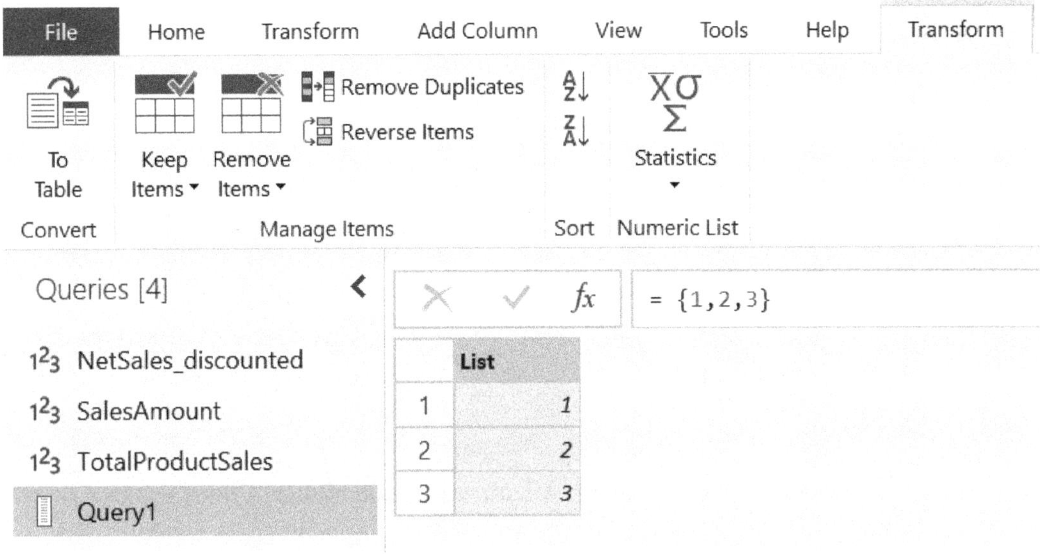

Figure 7.47 – List in Power Query UI

You can convert this list into a table or keep it as a list and use it as a parameter (please refer to the *Filtering with parameters* recipe in *Chapter 6, Optimizing Power Query Performance*). You can define a series of values and custom logic by exploring all of M's possibilities.

You have seen how you can write *free* code without depending necessarily on a data source and how you can derive values, integrating both calculations coming from data sources and custom logic defined natively in Power Query.

Creating tables in M

Thanks to M language, you can create complex tables from scratch without necessarily defining them inside an external data source and then importing them in Power Query. One common example is the definition of a list of dates that can then be customized.

Getting ready

In this recipe, you only need to have Power BI Desktop running on your PC.

How to do it...

Once you open your Power BI Desktop application, you are ready to perform the following steps:

1. Click on **Get Data** and select the **Blank query** connector.

2. The Power Query UI will pop up and you will see an empty query with its default name as **Query1**:

Figure 7.48 – Blank query

3. Browse to the **Home** tab and click on **Advanced Editor**.

   We want to create a list of dates that starts at `01/01/2011` and ends on the current date. In order to define this logic, we have to use the `List.Dates` expression and its syntax is as follows:

   ```
   List.Dates(start as date, count as number, step as
   duration) as list
   ```

   This expression consists of the following parts:

   a) `start`: starting date.

   b) `count`: values that refer to the number of dates to be retrieved.

   c) `step`: type of increment (for example, daily, monthly, and so on).

   Given our requirements, we have to obtain the following:

   a) `start`: `#date(2011,01,01)`.

   b) `count`: `Duration.Days (DateTime.Date(DateTime.LocalNow())`
   `- #date(2011,01,01))`, where we performed a subtraction to calculate the number of days between the current date (`DateTime.Date(DateTime.LocalNow())`) and the date defined as our start date.

   c) `step`: `#duration(1,0,0,0)`, where we defined one day as the value to increment the list of dates by.

In this way, we will get a list of dates that will look like the following:

`01/01/2011`

`02/01/2011`

...

`Current date`

The code to be inserted in the Advanced Editor will look like the following example. Insert this, and then click on **Done**:

```
let
    datelist = List.Dates(#date(2011,01,01),Duration.Days
    (DateTime.Date(DateTime.LocalNow()) - #date(2011,01,01)
    ), #duration(1,0,0,0))
in
        datelist
```

4. You can see how a list of dates has been created with the logic defined in the previous steps:

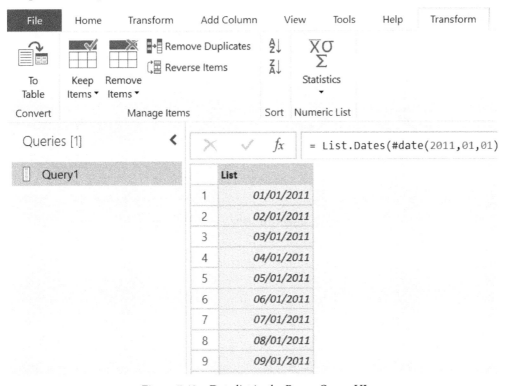

Figure 7.49 – Date list in the Power Query UI

5. Browse to the **Transform** section under **List Tools** and click on **To Table**:

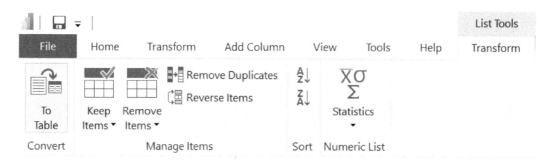

Figure 7.50 – To Table button

6. The **To Table** window will appear, and you will find that **None** as **Select or enter delimiter** and **Show as errors** as **How to handle extra columns** are selected by default. Leave it as it is and click on **OK**:

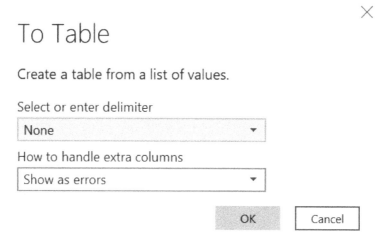

Figure 7.51 – To Table window

7. You will see that the list is converted to a table, and you will be able to customize your date table. First, change the column type to **Date** and rename the column to Date:

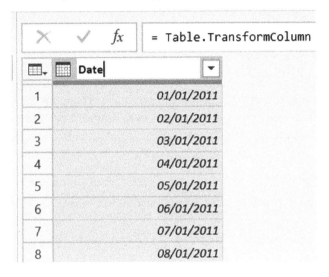

Figure 7.52 – Changed type and renamed to Date

8. Now you can browse to the **Add Column** tab, click on **Date**, and select which columns to add referencing to the Date column. You can add **Year**, **Month**, or **Quarter** and build your Date table:

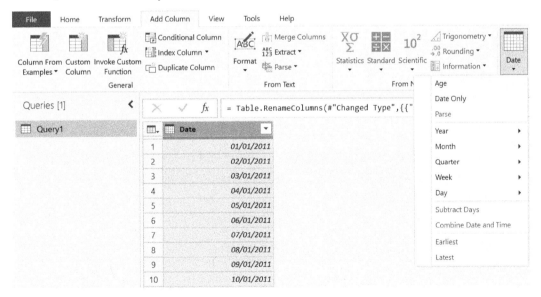

Figure 7.53 – Converting list into a table

Every time that you refresh your data, you will have an updated `Date` table that can be used for further purposes. This is an example of how you can leverage some `Date` expressions in M language, applying intuitive logic and getting the most out of this tool.

Leveraging M – tips and tricks

Using M code and editing existing queries from the Advanced Editor implies paying attention to some general rules in order to avoid common errors. In this recipe, we will discover some tips and tricks to keep in mind when editing queries and using M code on them.

Getting ready

In this recipe, you need to download the `FactInternetSales.csv` file.

In this example, we will refer to the `C:\Data` folder.

How to do it...

Once you open your Power BI Desktop application, you are ready to perform the following steps:

1. Click on **Get Data** and select the **Text/CSV** connector.

2. Browse to your local folder where you downloaded the `FactInternetSales.csv` file and open it. The following window, with a preview of the data, will pop up. Click on **Transform Data**.

3. Rename the `ProductKey` column to `ProductKeyCode`:

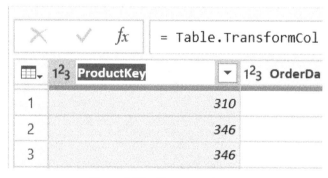

Figure 7.54 – Rename column

4. Click on **Choose Columns** and flag **ProductKeyCode**, **OrderQuantity**, **SalesAmount**, and **OrderDate**:

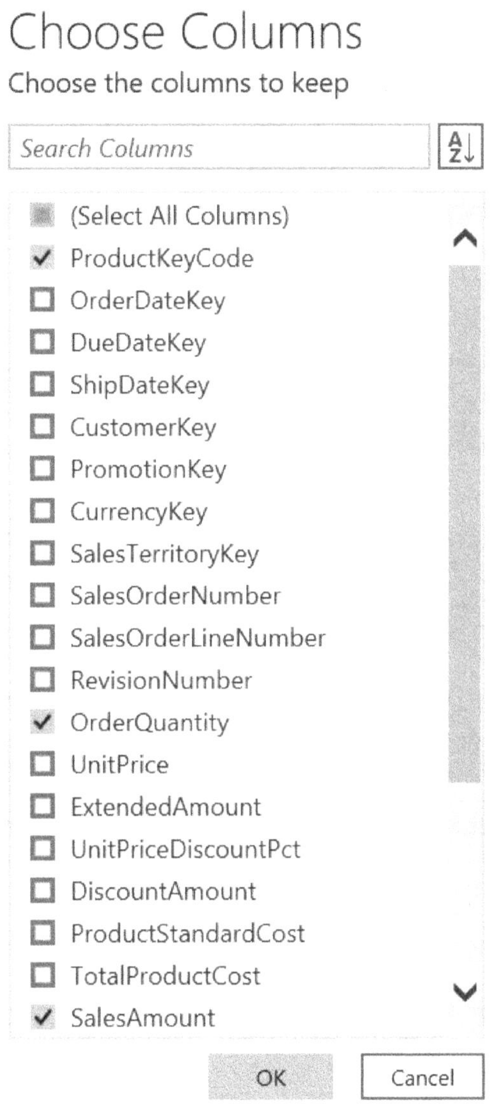

Figure 7.55 – Choose columns

5. Now open the **Advanced Editor** and, at line 5, edit the renamed value, and instead of ProductKeyCode write ProductKey_Code:

```
1  let
2      Source = Csv.Document(File.Contents("C:\Data\FactInternetSales.csv"),[Delimiter=",", Columns=26, Encoding=65001,
           QuoteStyle=QuoteStyle.None]),
3      #"Promoted Headers" = Table.PromoteHeaders(Source, [PromoteAllScalars=true]),
4      #"Changed Type" = Table.TransformColumnTypes(#"Promoted Headers",{{"ProductKey", Int64.Type}, {"OrderDateKey", Int64.Type},
           {"DueDateKey", Int64.Type}, {"ShipDateKey", Int64.Type}, {"CustomerKey", Int64.Type}, {"PromotionKey", Int64.Type},
           {"CurrencyKey", Int64.Type}, {"SalesTerritoryKey", Int64.Type}, {"SalesOrderNumber", type text}, {"SalesOrderLineNumber",
           Int64.Type}, {"RevisionNumber", Int64.Type}, {"OrderQuantity", Int64.Type}, {"UnitPrice", type number}, {"ExtendedAmount",
           type number}, {"UnitPriceDiscountPct", Int64.Type}, {"DiscountAmount", Int64.Type}, {"ProductStandardCost", type number},
           {"TotalProductCost", type number}, {"SalesAmount", type number}, {"TaxAmt", type number}, {"Freight", type number},
           {"CarrierTrackingNumber", type text}, {"CustomerPONumber", type text}, {"OrderDate", type datetime}, {"DueDate", type
           datetime}, {"ShipDate", type datetime}}),
5      #"Renamed Columns" = Table.RenameColumns(#"Changed Type",{{"ProductKey", "ProductKeyCode"}}),
6      #"Removed Other Columns" = Table.SelectColumns(#"Renamed Columns",{"ProductKeyCode", "OrderQuantity", "SalesAmount", "OrderDate"}
           )
7  in
8      #"Removed Other Columns"
```

Figure 7.56 – Edit column name in Advanced Editor

After renaming the column, click on **Done**. You can see that No syntax errors were detected and everything seems to be OK.

6. You won't see any data, but you will incur an error that looks like the one in the following screenshot:

Figure 7.57 – Error displayed

This error refers to the fact that the **Choose Columns** step expects a column called ProductKeyCode, but at *Step* 6 we renamed it ProductKey_Code without changing it in the step that followed.

7. Once again, open the **Advanced Editor**. In *line 6*, rename the ProductKeyCode column to ProductKey_Code and click on **Done**:

```
5      #"Renamed Columns" = Table.RenameColumns(#"Changed Type",{{"ProductKey", "ProductKey_Code"}}),
6      #"Removed Other Columns" = Table.SelectColumns(#"Renamed Columns",{"ProductKey_Code", "OrderQuantity", "SalesAmount",
           "OrderDate"})
7  in
8      #"Removed Other Columns"
```

✓ No syntax errors have been detected.

Done   Cancel

Figure 7.58 – Update column name in the Removed Other Columns step

8. You will now see that the data is displayed correctly:

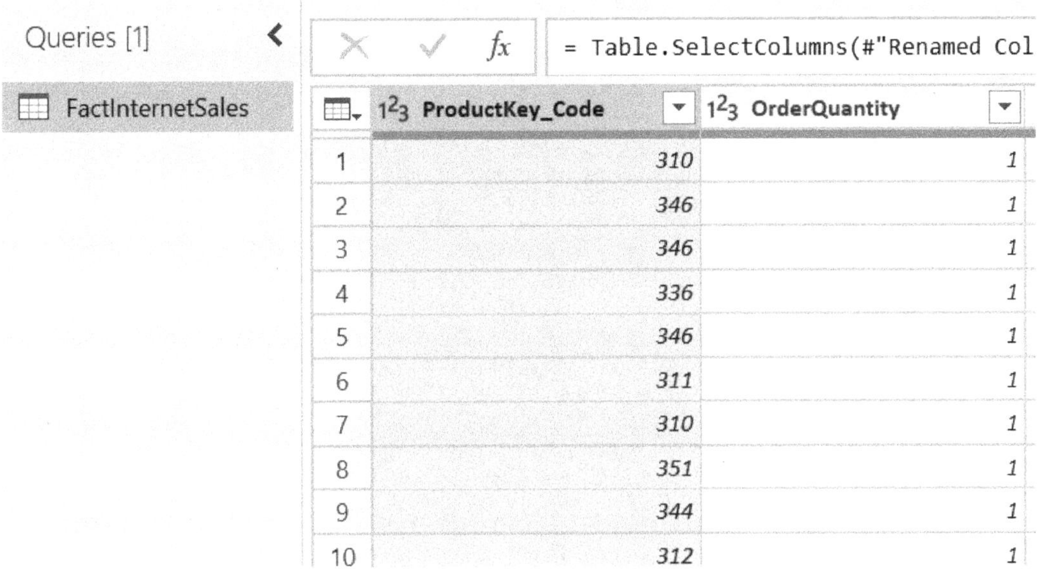

Figure 7.59 – Corrected query output

This example suggests that you have to pay attention when you edit existing steps and be aware of the potential for errors when you modify the names in the steps that follow. Also, pay attention to separate the different steps with a comma (,).

8
Adding Value to Your Data

You have the chance to connect to your data and create and transform it as you want thanks to a wide range of options explored in the previous chapters. Moreover, **Power Query** offers the chance to add data and enrich it with additional columns or define some functions to retrieve data. By adding columns, you can define your own customized logic in a few steps and leverage the UI or **M code** expressions, which represents the language behind the scenes of Power Query. M code expressions can be used to build functions and define input values and programmatically retrieve a defined output in order to simplify the entire transformation process.

In this chapter, you will explore how you can add data as new columns based on a pattern or a logic of existing data enriching it with valuable information and using a set of transformations.

In this chapter, we will cover the following recipes:

- Adding columns from examples
- Adding conditional columns
- Adding custom columns
- Invoking custom functions
- Clustering values

Technical requirements

For this chapter, you will be using the following:

- **Power BI Desktop**: `https://www.microsoft.com/en-us/download/details.aspx?id=58494`

The minimum requirements for installation are as follows:

- .NET Framework 4.6 (Gateway release August 2019 and earlier)

- .NET Framework 4.7.2 (Gateway release September 2019 and later)

- A 64-bit version of Windows 8 or a 64-bit version of Windows Server 2012 R2 with current TLS 1.2 and cipher suites

- 4 GB disk space for performance monitoring logs

You can find the data resources referred to in this chapter at `https://github.com/PacktPublishing/Power-Query-Cookbook/tree/main/Chapter08`.

Adding columns from examples

We often need to add new columns based on the structure or values of an already existing column or set of columns. Imagine you want to extract information from an existing column quickly or to concatenate some values by typing an example and then apply an underlying rule to all the values of that column. These scenarios can be easily achieved by building new content by adding columns from examples. In this recipe, we will see how to best leverage this feature.

Getting ready

For this recipe, you need to download the `FactInternetSales` CSV file.

In this example, we will refer to the `C:\Data` folder.

How to do it...

Once you open your Power BI Desktop application, you are ready to perform the following steps:

1. Click on **Get data** and select the **Text/CSV** connector.

2. Browse to your local folder where you downloaded the `FactInternetSales` CSV file and open it. The following window with a preview of the data will pop up; click on **Transform Data**:

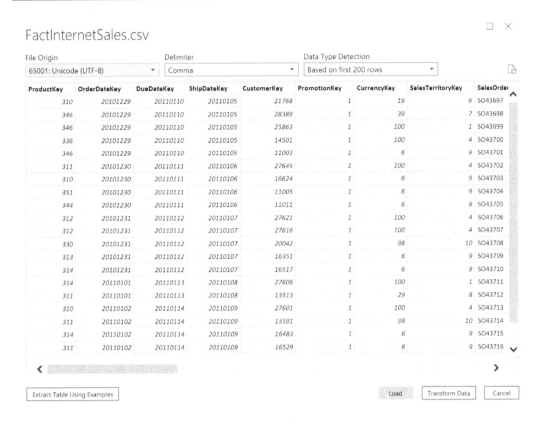

Figure 8.1 – CSV data preview

3. Select the `OrderDate` column, browse to the **Add Column** tab, and click on **Column From Examples** and then **From Selection**, as shown in the following screenshot:

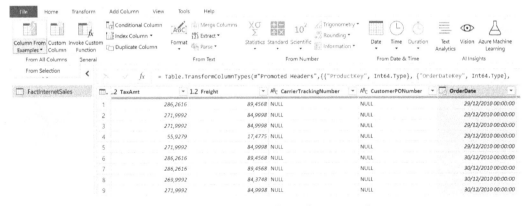

Figure 8.2 – Adding a column from examples

4. The usual Power Query interface will change and you will enter **Add Column From Examples** mode, where you will have the previously selected column flagged, while on the right side, you will have an empty column, **Column1**, ready to be created.

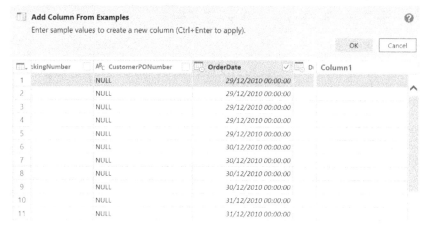

Figure 8.3 – Add Column From Examples section

5. Double-click on the first empty cell under **Column1** and you will see a dropdown appearing with some suggestions with values that can be extracted from the selected column.

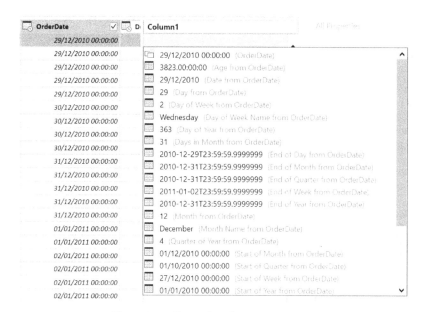

Figure 8.4 – Example columns suggestion

6. In this case, we want to extract the month expressed in letters and the year expressed in numbers in order to convert the value `29/12/2010 00:00:00` into `December-2010`. Type into the first cell the value you want to obtain, as in the following example, and press *Enter* on your keyboard:

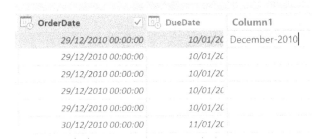

Figure 8.5 – Writing first column input

7. Then, go to row `15` and type `January-2011`, as in the following screenshot:

	ᴬᴮc CustomerPONumber	OrderDate	DueDate	Column1	⌒
1	NULL	29/12/2010 00:00:00	10/01/2C	December-2010	
2	NULL	29/12/2010 00:00:00	10/01/2C		
3	NULL	29/12/2010 00:00:00	10/01/2C		
4	NULL	29/12/2010 00:00:00	10/01/2C		
5	NULL	29/12/2010 00:00:00	10/01/2C		
6	NULL	30/12/2010 00:00:00	11/01/2C		
7	NULL	30/12/2010 00:00:00	11/01/2C		
8	NULL	30/12/2010 00:00:00	11/01/2C		
9	NULL	30/12/2010 00:00:00	11/01/2C		
10	NULL	31/12/2010 00:00:00	12/01/2C		
11	NULL	31/12/2010 00:00:00	12/01/2C		
12	NULL	31/12/2010 00:00:00	12/01/2C		
13	NULL	31/12/2010 00:00:00	12/01/2C		
14	NULL	31/12/2010 00:00:00	12/01/2C		
15	NULL	01/01/2011 00:00:00	13/01/2C	January-2011	
16	NULL	01/01/2011 00:00:00	13/01/2C		

Figure 8.6 – Writing second column input

8. Press the *Enter* key or click on another row and observe how the other rows are filled in following the two examples you have written.

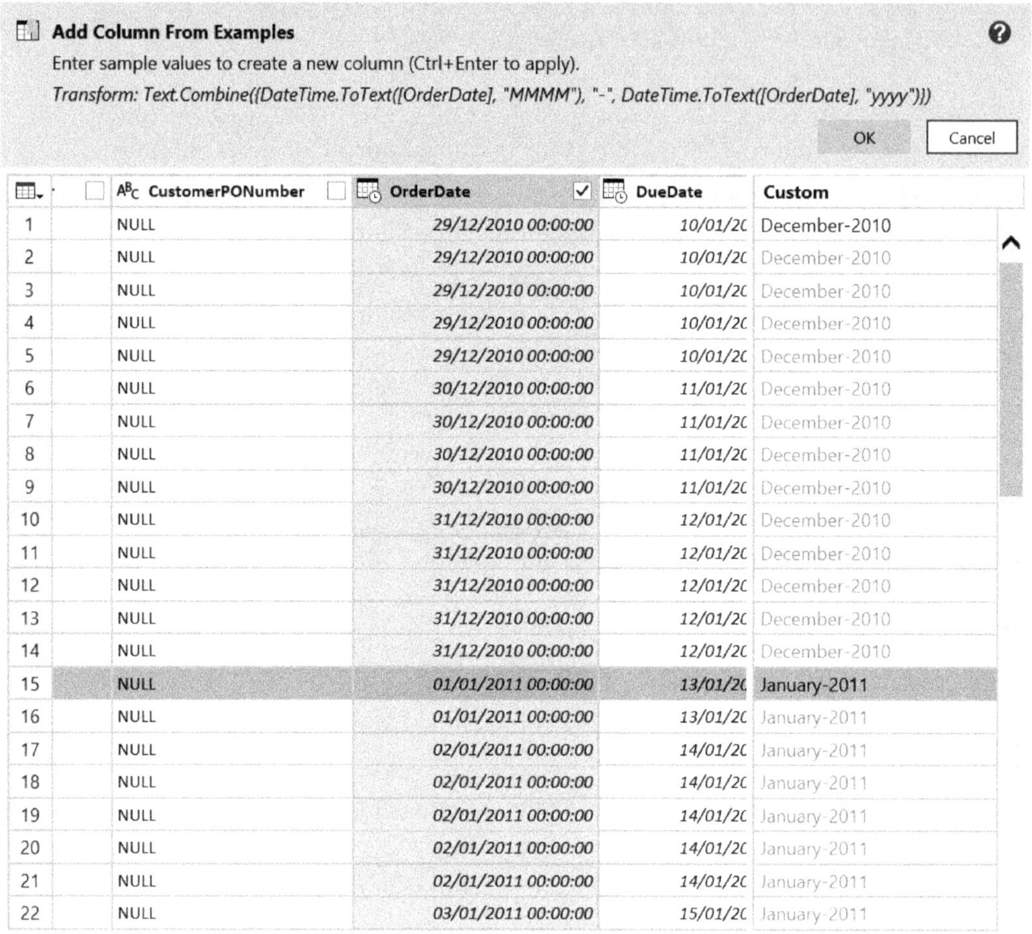

Add Column From Examples

Enter sample values to create a new column (Ctrl+Enter to apply).

Transform: *Text.Combine({DateTime.ToText([OrderDate], "MMMM"), "-", DateTime.ToText([OrderDate], "yyyy")})*

		ABC CustomerPONumber		OrderDate	✓	DueDate	Custom
1		NULL		29/12/2010 00:00:00		10/01/20	December-2010
2		NULL		29/12/2010 00:00:00		10/01/20	December-2010
3		NULL		29/12/2010 00:00:00		10/01/20	December-2010
4		NULL		29/12/2010 00:00:00		10/01/20	December-2010
5		NULL		29/12/2010 00:00:00		10/01/20	December-2010
6		NULL		30/12/2010 00:00:00		11/01/20	December-2010
7		NULL		30/12/2010 00:00:00		11/01/20	December-2010
8		NULL		30/12/2010 00:00:00		11/01/20	December-2010
9		NULL		30/12/2010 00:00:00		11/01/20	December-2010
10		NULL		31/12/2010 00:00:00		12/01/20	December-2010
11		NULL		31/12/2010 00:00:00		12/01/20	December-2010
12		NULL		31/12/2010 00:00:00		12/01/20	December-2010
13		NULL		31/12/2010 00:00:00		12/01/20	December-2010
14		NULL		31/12/2010 00:00:00		12/01/20	December-2010
15		NULL		01/01/2011 00:00:00		13/01/20	January-2011
16		NULL		01/01/2011 00:00:00		13/01/20	January-2011
17		NULL		02/01/2011 00:00:00		14/01/20	January-2011
18		NULL		02/01/2011 00:00:00		14/01/20	January-2011
19		NULL		02/01/2011 00:00:00		14/01/20	January-2011
20		NULL		02/01/2011 00:00:00		14/01/20	January-2011
21		NULL		02/01/2011 00:00:00		14/01/20	January-2011
22		NULL		03/01/2011 00:00:00		15/01/20	January-2011

Figure 8.7 – Input autofill

9. Rename the column that has the temporary name Custom (because you have been creating a custom column) and call it OrderDate-MthYear.

OrderDate-MthYear

)1/2(December-2010

Figure 8.8 – Column name

10. Observe the preview of the M code at the top of the section and the functions that were applied. At the end, click on **OK**.

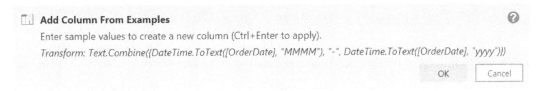

Figure 8.9 – M code generated by the column creation

11. You can see the new column now and the newly added step in the **APPLIED STEPS** section, **Added Custom Column**.

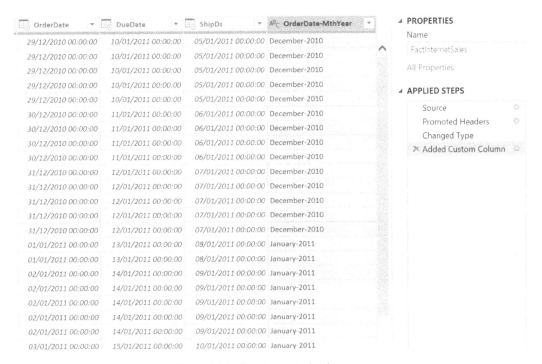

Figure 8.10 – Newly created column step

This feature allows you to quickly create new columns by just entering the example you have in mind or a few inputs to see what the suggestion of the tool will be.

Imagine you want to concatenate values from multiple columns and enrich your query with more data. You can easily achieve this with the same functionality by following the next example:

1. Select the `SalesOrderNumber` and `SalesOrderLineNumber` columns, browse to the **Add Column** tab, and click on **Column From Examples** and then **From Selection**, as shown in the following screenshot:

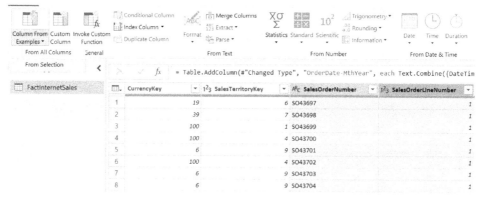

Figure 8.11 – Columns selection

2. Double-click on the first cell under `Column1` and select **SO43697** and you will get the first part of the new value.

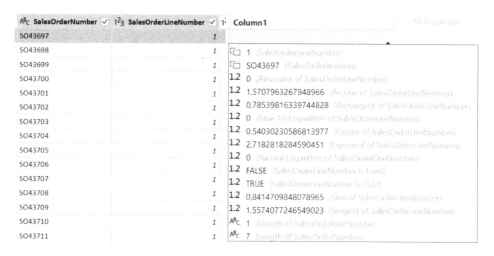

Figure 8.12 – New column from examples

3. You can see the temporary values of this new column.

Figure 8.13 – Entering the first input

4. Now, click again on the first cell and let's add the second part of the value.
    Enter -SOLN1 to perform a concatenation between the following values -
    SalesOrderNumber, added in the previous step, then a separator (-), then the
    initials of SalesOrderLineNumber (SOLN), and at the end, the value of it (1), as
    shown in the following screenshot:

A^BC SalesOrderNumber	✓ 1²₃ SalesOrderLineNumber	✓ 1· Merged
SO43697	1	SO43697-SOLN1
SO43698	1	SO43698-SOLN1
SO43699	1	SO43699-SOLN1
SO43700	1	SO43700-SOLN1
SO43701	1	SO43701-SOLN1
SO43702	1	SO43702-SOLN1
SO43703	1	SO43703-SOLN1
SO43704	1	SO43704-SOLN1
SO43705	1	SO43705-SOLN1
SO43706	1	SO43706-SOLN1
SO43707	1	SO43707-SOLN1
SO43708	1	SO43708-SOLN1
SO43709	1	SO43709-SOLN1
SO43710	1	SO43710-SOLN1
SO43711	1	SO43711-SOLN1
SO43712	1	SO43712-SOLN1
SO43713	1	SO43713-SOLN1
SO43714	1	SO43714-SOLN1
SO43715	1	SO43715-SOLN1
SO43716	1	SO43716-SOLN1
SO43717	1	SO43717-SOLN1
SO43718	1	SO43718-SOLN1
SO43719	1	SO43719-SOLN1

Figure 8.14 – Second input within the same cell

5. See how the concatenation was performed by having a look at the M code applied, and then click on **OK**.

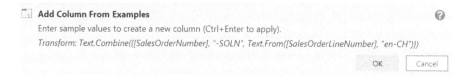

Figure 8.15 – M code generated by the step

6. At the end, rename the column `SalesOrderFullCode`.

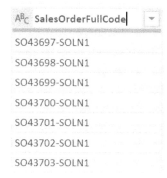

Figure 8.16 – Column renaming

Adding columns from examples offers different options when you do not want to code directly with M or when you know exactly what the desired output is but you are not sure which transformation options in Power Query to use.

Adding conditional columns

You can also enrich your queries by adding columns with `if…then` logic and applying some conditions based on existing columns. These are called **conditional columns** and this feature allows you to implement conditional expressions with an intuitive interface. In this recipe, you will see how it is easy to apply custom flags based on the values of existing columns.

Getting ready

For this recipe, you need to download the `FactInternetSales` CSV file.

In this example, we will refer to the `C:\Data` folder.

How to do it...

Once you open your Power BI Desktop application, you are ready to perform the following steps:

1. Click on **Get data** and select the **Text/CSV** connector.

2. Browse to your local folder where you downloaded the `FactInternetSales` CSV file and open it. The following window with a preview of the data will pop up; click on **Transform Data**:

ProductKey	OrderDateKey	DueDateKey	ShipDateKey	CustomerKey	PromotionKey	CurrencyKey	SalesTerritoryKey	SalesOrder
310	20101229	20110110	20110105	21768	1	19	6	SO43697
346	20101229	20110110	20110105	28389	1	39	7	SO43698
346	20101229	20110110	20110105	25863	1	100	1	SO43699
336	20101229	20110110	20110105	14501	1	100	4	SO43700
346	20101229	20110110	20110105	11003	1	6	9	SO43701
311	20101230	20110111	20110106	27645	1	100	4	SO43702
310	20101230	20110111	20110106	16624	1	6	9	SO43703
351	20101230	20110111	20110106	11005	1	6	9	SO43704
344	20101230	20110111	20110106	11011	1	6	9	SO43705
312	20101231	20110112	20110107	27621	1	100	4	SO43706
312	20101231	20110112	20110107	27616	1	100	4	SO43707
330	20101231	20110112	20110107	20042	1	98	10	SO43708
313	20101231	20110112	20110107	16351	1	6	9	SO43709
314	20101231	20110112	20110107	16517	1	6	9	SO43710
314	20110101	20110113	20110108	27606	1	100	1	SO43711
311	20110101	20110113	20110108	13513	1	29	8	SO43712
310	20110102	20110114	20110109	27601	1	100	4	SO43713
311	20110102	20110114	20110109	13591	1	98	10	SO43714
314	20110102	20110114	20110109	16483	1	6	9	SO43715
311	20110102	20110114	20110109	16529	1	6	9	SO43716

Extract Table Using Examples   Load   Transform Data   Cancel

Figure 8.17 – CSV data preview

3. Browse to the **Add Column** tab and click on **Conditional Column**.

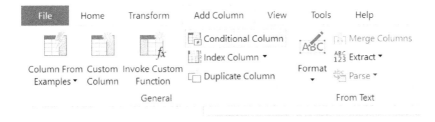

Figure 8.18 – Conditional Column button

4. The **Add Conditional Column** window will pop up, as in the following screenshot:

Figure 8.19 – Add Conditional Column window

5. Name the new column `PriceLevel` and start creating the first condition by selecting **UnitPrice** from the first dropdown after **If**.

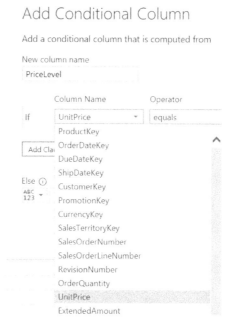

Figure 8.20 – Defining the condition

6. Then select **is greater than** from the **Operator** dropdown and enter 3000 for **Value** and High for **Output,** as in the following example, in order to label unit prices higher than 3000 with the High flag:

Figure 8.21 – Defining Value and Output

7. Then, click on **Add Clause** in order to add another two conditions and fill them in, as in the following screenshot. At the end, click on **OK**.

Add Conditional Column

Add a conditional column that is computed from the other columns or values.

New column name

PriceLevel

	Column Name	Operator	Value	Output	
If	UnitPrice	is greater than	ABC 123 3000	Then ABC 123 High	...
Else If	UnitPrice	is greater than	ABC 123 2000	Then ABC 123 Medium-High	
Else If	UnitPrice	is greater than	ABC 123 1000	Then ABC 123 Medium	

Add Clause

Else

ABC 123 Low

OK   Cancel

Figure 8.22 – Defining multiple conditions

With this example, we are stating that products with `UnitPrice` higher than 3000 have to be labeled as `High`, between 2000 and 3000 as `Medium-High`, and between 1000 and 2000 as `Medium`, and if any of these conditions are not met, the label should be `Low` (defined by the input in the **Else** section on the bottom left in the preceding figure).

Figure 8.23 – New column created

The same `PriceLevel` column output could be achieved by setting up the conditions as follows:

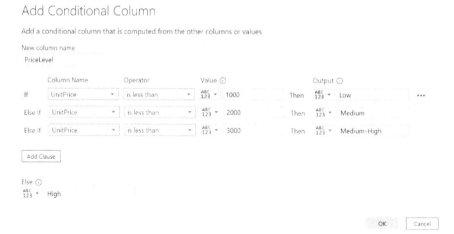

Figure 8.24 – Defining a different order

Moreover, you can use dynamic values both as input values and outputs by using column values or parameters.

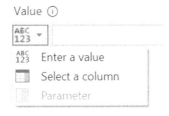

Figure 8.25 – Different value options

It is important to correctly define the order of the conditions. Applying it to our recipe will work as follows:

1. All unit prices higher than 3000 will be flagged as High since these rows satisfy that first condition.

2. Then, all prices higher than 2000, except ones already flagged by the first condition, will be flagged as Medium-High.

3. Then, all prices higher than 1000, except ones already flagged by the first and second conditions, will be flagged as Medium.

4. All other prices will be flagged as Low.

By paying attention to the order and selecting the right value, you can use conditional columns to create and enrich your data by applying conditions to different data types and defining custom outputs.

Adding custom columns

Once you have become more confident with Power Query M code, you can also enrich data content by writing formulas and expressions directly thanks to the custom columns feature. In this recipe, you will see an example of how to browse this section and create columns with calculations that are not available in the form of built-in features in the Power Query UI.

Getting ready

For this recipe, you need to download the FactInternetSales CSV file.

In this example, we will refer to the C:\Data folder.

How to do it...

Once you open your Power BI Desktop application, you are ready to perform the following steps:

1. Click on **Get data** and select the **Text/CSV** connector.

2. Browse to your local folder where you downloaded the `FactInternetSales` CSV file and open it. The following window with a preview of the data will pop up; click on **Transform Data**:

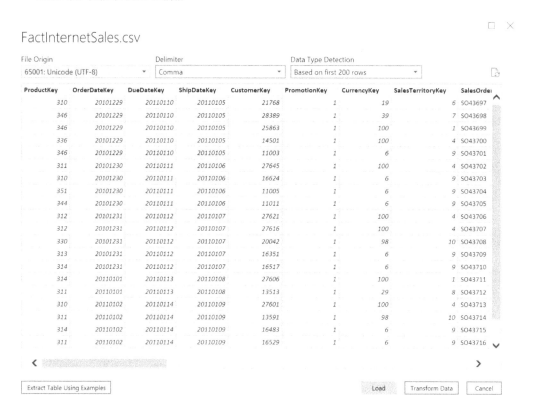

Figure 8.26 – CSV data preview

3. Imagine you want to calculate how many days have passed between two dates, in this case, between the ship and order dates. You can do that easily by leveraging the `Duration.Days()` M function, which can be applied in the **Custom Column** section. Browse to the **Add Column** tab and click on the **Custom Column** button.

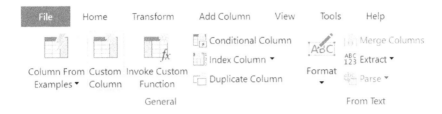

Figure 8.27 – Custom Column button

4. The **Custom Column** window will pop up and you will have the chance to create an output column from an M expression by referencing other existing columns.

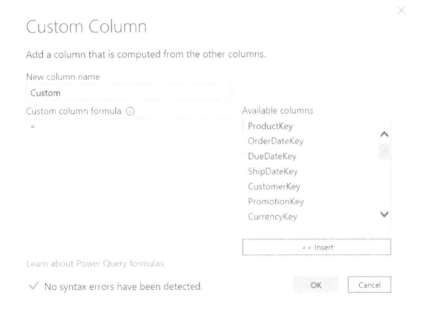

Figure 8.28 – Custom Column window

5. Assign a new column name, in this case, `NrShippingDays`, and start typing the `Duration.Days` function. You will see IntelliSense working live when presented with function options.

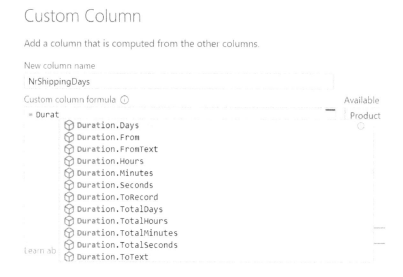

Figure 8.29 – IntelliSense formula engine

6. Once you have typed the function and opened the first bracket, you do not have to necessarily write down the names of the columns, but you can use the section on the right with the columns list to directly insert the query you need rather than writing it from scratch, as shown in the following screenshot:

Figure 8.30 – Formula definition

Once you have clicked on < < **Insert**, you will see the column appearing in the **Custom column formula** section.

New column name

NrShippingDays

Custom column formula ⓘ

= Duration.Days([ShipDate])

Learn about Power Query formulas

! Token RightParen expected.   Show error

Figure 8.31 – Formula definition

Do not worry about the red squiggly line because it appears since the format is temporarily not right (there are missing brackets essentially).

7.   Complete the function by adding a minus (-) and the date to be subtracted, OrderDate, and then close the brackets and click on **OK**.

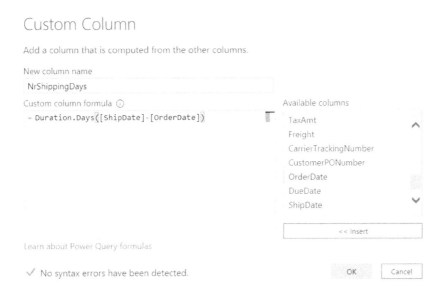

Custom Column

Add a column that is computed from the other columns.

New column name

NrShippingDays

Custom column formula ⓘ   Available columns

= Duration.Days([ShipDate]-[OrderDate])   TaxAmt
                                                 Freight
                                                 CarrierTrackingNumber
                                                 CustomerPONumber
                                                 OrderDate
                                                 DueDate
                                                 ShipDate

                                                 << Insert

Learn about Power Query formulas

✓ No syntax errors have been detected.   OK   Cancel

Figure 8.32 – Adding existing columns

You can notice how the message in the bottom left confirms **No syntax errors have been detected**.

8.   You can now see the new column created.

OrderDate	DueDate	ShipDate	NrShippingDays
29/12/2010 00:00:00	10/01/2011 00:00:00	05/01/2011 00:00:00	7
29/12/2010 00:00:00	10/01/2011 00:00:00	05/01/2011 00:00:00	7
29/12/2010 00:00:00	10/01/2011 00:00:00	05/01/2011 00:00:00	7
29/12/2010 00:00:00	10/01/2011 00:00:00	05/01/2011 00:00:00	7
29/12/2010 00:00:00	10/01/2011 00:00:00	05/01/2011 00:00:00	7
30/12/2010 00:00:00	11/01/2011 00:00:00	06/01/2011 00:00:00	7
30/12/2010 00:00:00	11/01/2011 00:00:00	06/01/2011 00:00:00	7
30/12/2010 00:00:00	11/01/2011 00:00:00	06/01/2011 00:00:00	7
30/12/2010 00:00:00	11/01/2011 00:00:00	06/01/2011 00:00:00	7
31/12/2010 00:00:00	12/01/2011 00:00:00	07/01/2011 00:00:00	7
31/12/2010 00:00:00	12/01/2011 00:00:00	07/01/2011 00:00:00	7
31/12/2010 00:00:00	12/01/2011 00:00:00	07/01/2011 00:00:00	7

Figure 8.33 – New column added

You can also create more dynamic calculations. Imagine you want to calculate the difference between today's date and the order date or any other date value. You would have to replace [ShipDate] in the previous formula with the DateTime.LocalNow() function, as shown:

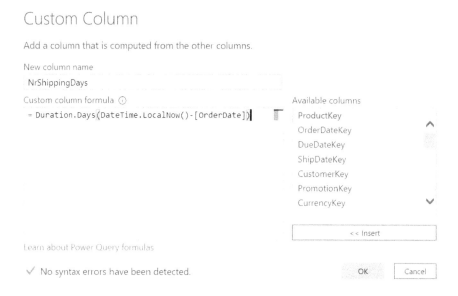

Figure 8.34 – Second formula definition

You can see the newly created `DaysFromCurrentDate` column.

OrderDate	DueDate	ShipDate	DaysFromCurrentDate
29/12/2010 00:00:00	10/01/2011 00:00:00	05/01/2011 00:00:00	3824
29/12/2010 00:00:00	10/01/2011 00:00:00	05/01/2011 00:00:00	3824
29/12/2010 00:00:00	10/01/2011 00:00:00	05/01/2011 00:00:00	3824
29/12/2010 00:00:00	10/01/2011 00:00:00	05/01/2011 00:00:00	3824
29/12/2010 00:00:00	10/01/2011 00:00:00	05/01/2011 00:00:00	3824
30/12/2010 00:00:00	11/01/2011 00:00:00	06/01/2011 00:00:00	3823
30/12/2010 00:00:00	11/01/2011 00:00:00	06/01/2011 00:00:00	3823
30/12/2010 00:00:00	11/01/2011 00:00:00	06/01/2011 00:00:00	3823
30/12/2010 00:00:00	11/01/2011 00:00:00	06/01/2011 00:00:00	3823
31/12/2010 00:00:00	12/01/2011 00:00:00	07/01/2011 00:00:00	3822
31/12/2010 00:00:00	12/01/2011 00:00:00	07/01/2011 00:00:00	3822
31/12/2010 00:00:00	12/01/2011 00:00:00	07/01/2011 00:00:00	3822
31/12/2010 00:00:00	12/01/2011 00:00:00	07/01/2011 00:00:00	3822
31/12/2010 00:00:00	12/01/2011 00:00:00	07/01/2011 00:00:00	3822
01/01/2011 00:00:00	13/01/2011 00:00:00	08/01/2011 00:00:00	3821
01/01/2011 00:00:00	13/01/2011 00:00:00	08/01/2011 00:00:00	3821

Figure 8.35 – New column created

Every time you refresh your data, the number will update according to the current date.

You can see it is easy to create custom columns and the more you become confident with M functions, the easier it will be to create more complex content. You will be able to concatenate, perform complex calculations, and refer to parameters as you would do with M functions in the Advanced Editor.

Invoking custom functions

Power Query offers you the ability to enrich existing tables with additional columns in many different ways, as you have seen in previous recipes, but it also allows you to use **custom functions** defined as expressions that take some variables as inputs to return a result value. In this recipe, we will see how to create a function, define function parameters, and invoke that function to generate an output.

Getting ready

For this recipe, you need to download the `FactResellerSales` CSV file.

In this example, we will refer to the `C:\Data` folder.

How to do it...

Once you open your Power BI Desktop application, you are ready to perform the following steps:

1. Click on **Get data** and select the **Text/CSV** connector.

2. Browse to your local folder where you downloaded the FactResellerSales CSV file and open it. The following window with a preview of the data will pop up; click on **Transform Data**:

Figure 8.36 – CSV data preview

3. Now, right-click on the **Queries** pane space and click on **New Query** and then **Blank Query**.

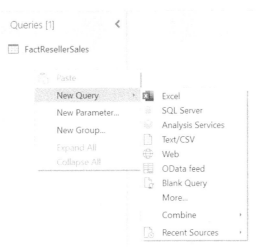

Figure 8.37 – Blank query creation

4. In our example, we will define a function to calculate the net sales amount when applying different discount values (if you have `SalesAmount` equal to `10`, you first apply a discount of 10%, and then you subtract the total cost from that discounted value). After having created a new blank query, browse to the **Home** tab and open **Advanced Editor**.

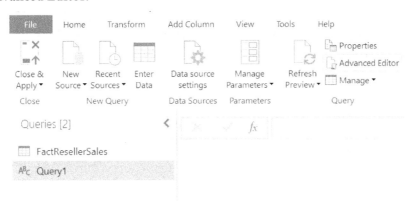

Figure 8.38 – Blank query display

5. A blank query contains the M code text you see in the following screenshot and here, we can define our own function:

Figure 8.39 – Blank query in the Advanced Editor

6. Enter the following code to create a function:

```
(OldSalesAmount as number, Discount as number, TotalCosts
as number) =>
let
   NetSales = OldSalesAmount - (OldSalesAmount * Discount
) - TotalCosts
in
    NetSales
```

The formula is divided into the following parts:

a) **Definition of input values**: OldSalesAmount, Discount, and TotalCosts. After you define these parameters, you will add the => expression to introduce the function and the subsequent part, which starts with let.

b) **Function definition**: Function formula to calculate NetSales, which is given by OldSalesAmount - (OldSalesAmount * Discount) - TotalCosts.

c) **Value returned**: This is introduced by the in clause.

Have a look at how it looks in the **Advanced Editor** and click on **Done**.

Figure 8.40 – Function definition in the Advanced Editor

7. See how the function appears in the Power Query UI.

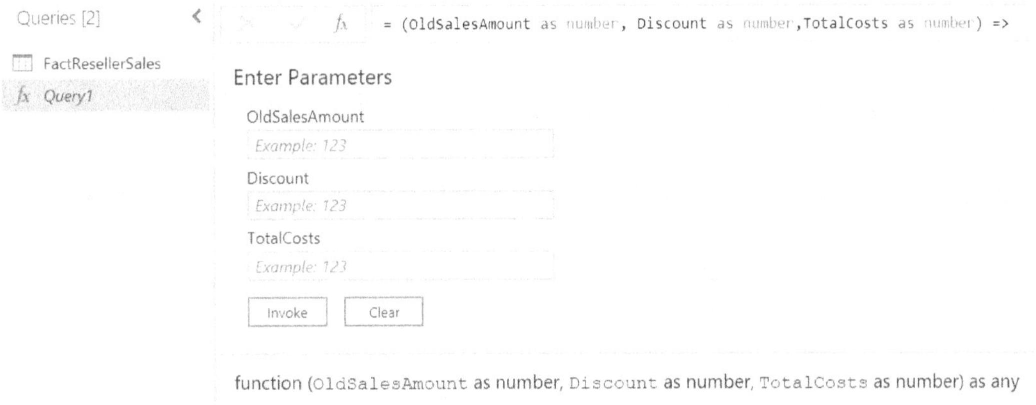

Figure 8.41 – Custom function parameters

Let's also try to manually input some values to see how it works. Enter 5 for **OldSalesAmount**, 0,1 for **Discount** (meaning 10%), and 3 for **TotalCosts**, and then click on the **Invoke** button.

Figure 8.42 – Enter Parameters

8. You will see the NetSales value when a 10% discount is applied as the output of the invoked function.

Figure 8.43 – Invoked Function

9. Now, let's delete **Invoked Function** under the **Queries** section on the left side of the UI and try to create an invoked custom function to apply this calculation to the **FactResellerSales** data.

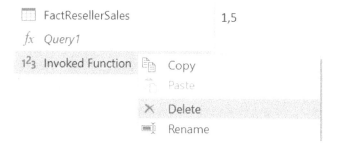

Figure 8.44 – Deleting Invoked Function

10. Rename the function query to `fxNetSales` to identify it easily as the function to calculate `NetSales`.

Figure 8.45 – Renaming a function

11. Now select the **FactResellerSales** query, browse to the **Add Column** tab, and click on **Invoke Custom Function** to apply the `fxNetSales` function, defining as inputs the columns from **FactResellerSales**.

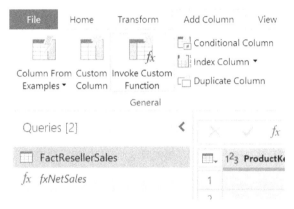

Figure 8.46 – Invoke Custom Function button

12. The **Invoke Custom Function** window will pop up and from here, you can define which function and what input variables to use. Name the new column `NetSales` and select the **fxNetSales** function from the **Function query** dropdown.

Figure 8.47 – Invoke Custom Function window

13. Select the **SalesAmount** column for the **OldSalesAmount** input, enter 0 , 1 for the **Discount** input, and then click on the input type icon and select **Column Name** for the **TotalCosts** input.

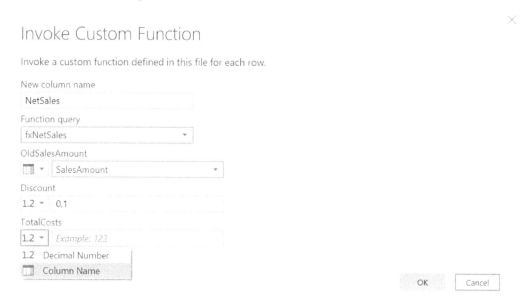

Figure 8.48 – Variables definition

14. Then select the **TotalProductCost** column as the input column for **TotalCosts**.

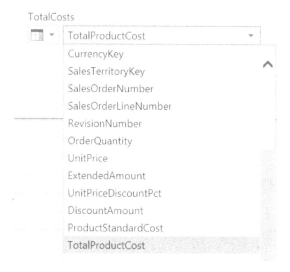

Figure 8.49 – Variable selection

15. After having defined the input variables, click on **OK**.

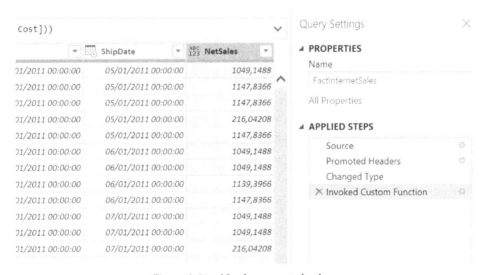

Figure 8.50 – Variables defined

16. A new column will be added to the query as the output of the function with the variables you have defined.

Figure 8.51 – Newly generated column

The example you saw is just one of many that you can perform using custom functions. The idea of this feature is to create something that can be reused with multiple queries and allow you to simplify, accelerate, and easily maintain your data transformation steps.

Clustering values

Data often comes from original data sources in many forms and you can end up having multiple variations of the same value where you need to have a unique value. In this case, you need a way to group and correct these values quickly, without creating complex rules or doing it manually. In this recipe, we will see how to leverage the clustering values feature, which enables you to automatically group data based on similarities thanks to an underlying algorithm.

Getting ready

For this recipe, you need to have access to the Power BI portal, for which a Power BI Pro license is needed. You also need to have access to a workspace.

How to do it...

After you log in to the Power BI portal, perform the following steps:

1. Browse to your workspace, click on **New**, and click on **Dataflow**.

Figure 8.52 – Creating a dataflow

2. Click on **Add new tables** in order to connect to a data source and access the Power Query online UI.

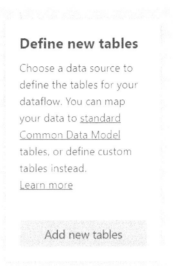

Figure 8.53 – Add new tables

3. Select the **Web API** connector and enter the following URL to connect to a CSV file loaded in the *Power Query Cookbook* GitHub repository, `https://github.com/PacktPublishing/Power-Query-Cookbook/blob/main/Chapter08/SalesData.csv`, and then click on **Next** to see the data preview.

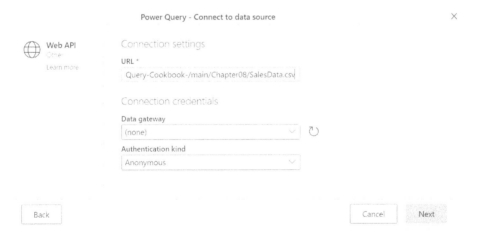

Figure 8.54 – Web API connector

4. You will now see a data preview with the **File origin**, **Delimiter**, and **Data type detection** options automatically detected. Click on **Transform data** to access the Power Query UI.

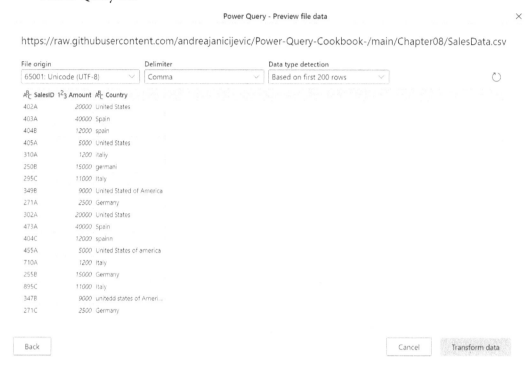

Figure 8.55 – Data preview

5. Now have a look at the Country column and see how countries' names are not spelled in the same way. Imagine you want to have homogeneous names and to correct capital letters and other spelling issues. In this case, we will use the **Cluster values** feature to create these clusters. Browse to **Add column** and click on the **Cluster values** button.

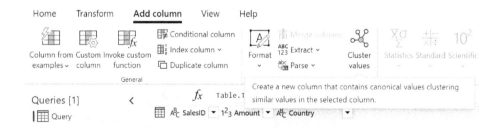

Figure 8.56 – Cluster values button

Select **Country** for **Column** and define a new column called Country_
corrected, which will contain the corrected values. Define a similarity threshold
of 0.7 (the default is 0.8), which indicates how two similar values should be to
be clustered together (for example, United States and United States of
America). **Ignore case** and **Group by combining text parts** will be flagged by
default. In this case, leave the default options flagged and also flag **Show similarity
scores**, and then click on **OK**.

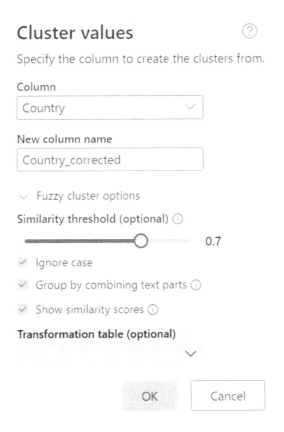

Figure 8.57 – Cluster values window

6. After you have clicked **OK**, you will see two newly added columns, one with the clustered values, `Country_corrected`, and one with score similarities, `Country_Country_corrected_Similarity`. You can observe how values were grouped according to similarities identified by the algorithm.

Figure 8.58 – Clusters column

With the **Cluster values** feature, you have the possibility to quickly correct values within a column by using a fuzzy matching algorithm built into Power Query online. You can think of different scenarios to correct your data and get the most out of this feature.

9
Performance Tuning with Power BI Dataflows

We already had the chance to see in detail how **Power Query** works in its Desktop version, where you can perform data preparation and transformations and save all of it in a **Power BI Desktop** file with the .pbix extension. But what if you would like to reuse Power Query transformations done by others but cannot retrieve the .pbix file? Or what if you want to store them somewhere accessible to multiple users? You can do that, thanks to the Power Query online version that is integrated with the **Power BI Dataflows** feature, accessible via the **Power BI Portal**.

In this chapter, we will see how to create, configure, and consume dataflows by exploring the following recipes:

- Using Power BI dataflows
- Centralizing ELT with dataflows
- Building dataflows with Power BI Premium capabilities
- Understanding dataflow best practices

Technical requirements

For this chapter, you will need the following:

- **Power BI Desktop**: `https://www.microsoft.com/en-us/download/details.aspx?id=58494`

- A **Power BI Pro** license and access to the `www.powerbi.com` portal

- A Power BI Premium or Power BI Premium Per User license

The minimum requirements for installation are as follows:

- .NET Framework 4.6 (Gateway release August 2019 or earlier)

- .NET Framework 4.7.2 (Gateway release September 2019 or later)

- A 64-bit version of Windows 8 or a 64-bit version of Windows Server 2012 R2 with current TLS 1.2 and cipher suites

- 4 GB of disk space for performance monitoring logs

You can find the data resources referred to in this chapter at the following link:

`https://github.com/PacktPublishing/Power-Query-Cookbook/tree/main/Chapter09`

Using Power BI dataflows

You can easily access **Power Query online** through Power BI dataflows, a feature you can find in the Power BI portal once you have logged in. You can create dataflows in Power BI workspaces and manage user permissions as you would do with reports and Power BI datasets. In this recipe, we will see how the Power BI dataflows feature works and what the main concepts to learn about are to get the most out of it.

Getting ready

For this recipe, you need to have access to the Power BI portal, for which a Power BI Pro license is needed.

How to do it...

After you log in to the Power BI portal, perform the following steps:

1. Create a new workspace, a place where you can collaborate with others, share and publish your content, and from where you can access Power Query online in the Power BI portal. You can do that by clicking on **Workspaces** and then on **Create a workspace** as shown in the screenshot:

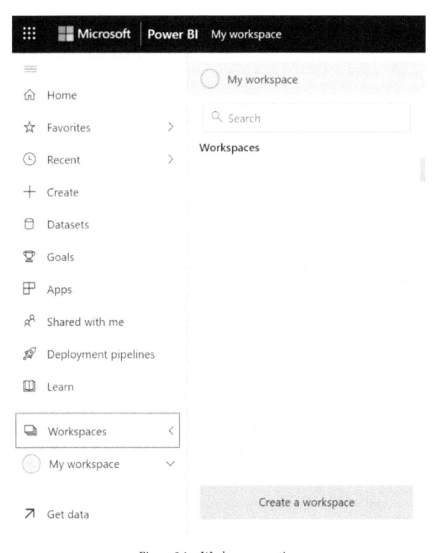

Figure 9.1 – Workspace creation

2. The following window will pop up on the right side of your browser, where you can enter your workspace name. In this recipe, we will call it `Power Query Cookbook` and enter this name in the **Workspace name** section:

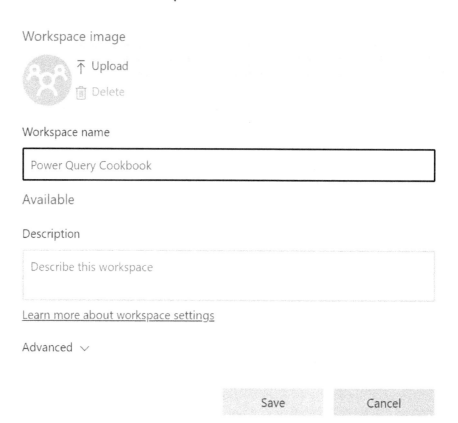

Figure 9.2 – Workspace creation

3.   Then expand the **Advanced** section, right above the **Save** button, and flag **Workspace admins** under **Contact list** (which means that only workspace admins will receive notifications about problems in the workspace) and choose a licensing option available in your environment. In this case, select **Pro** as in the following screenshot:

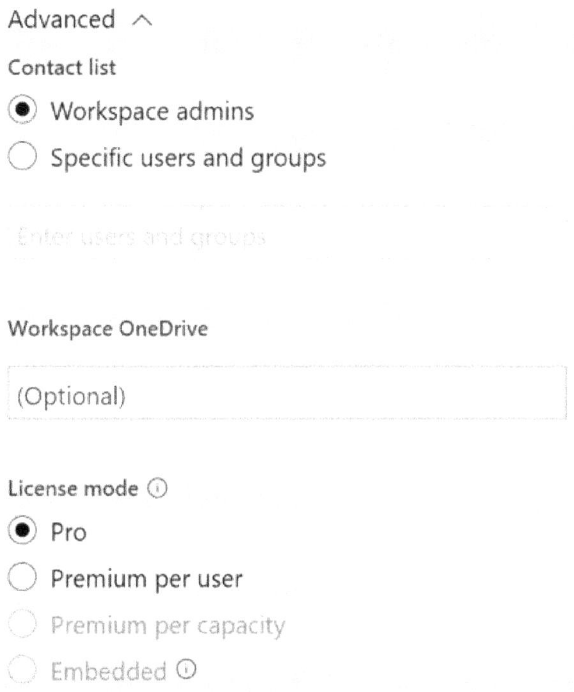

Advanced ⌃

Contact list
◉ Workspace admins
◯ Specific users and groups

Enter users and groups

Workspace OneDrive

(Optional)

License mode ⓘ
◉ Pro
◯ Premium per user
◯ Premium per capacity
◯ Embedded ⓘ

Figure 9.3 – Workspace creation Advanced section

4. Once you create the workspace, you will see this page from where you can start building content:

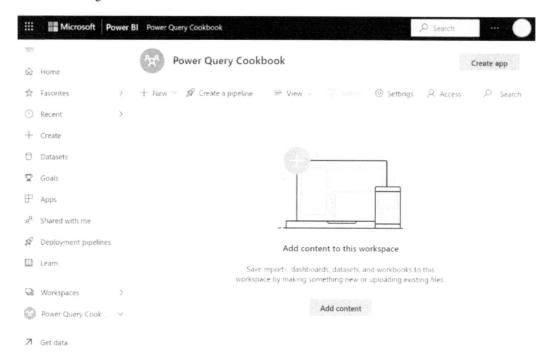

Figure 9.4 – Workspace page

5. Click on **New** and then click on **Dataflow** as in the following screenshot:

Power Query Cookbook
New
Create a pipeline
Report
Visualize your data
Paginated Report
Build a paginated report
Dashboard
Build a single-page data story
Dataset
Create a dataset to use in a report
Dataflow
Prep, clean, and transform data
Streaming dataset
Build visuals from real-time data
Upload a file
Open a .pbix, .rdl, .xlsx, or .csv in Po...

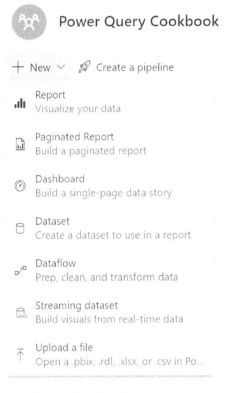

Figure 9.5 – Dataflow creation

6. The following options will be presented to start using the dataflow features:

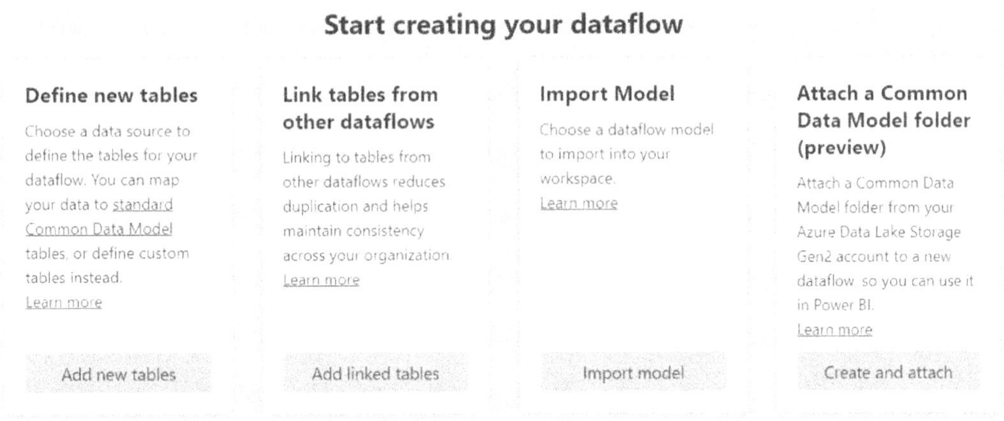

Start creating your dataflow
Define new tables
Choose a data source to define the tables for your dataflow. You can map your data to standard Common Data Model tables, or define custom tables instead.
Learn more
Add new tables
Link tables from other dataflows
Linking to tables from other dataflows reduces duplication and helps maintain consistency across your organization.
Learn more
Add linked tables
Import Model
Choose a dataflow model to import into your workspace.
Learn more
Import model
Attach a Common Data Model folder (preview)
Attach a Common Data Model folder from your Azure Data Lake Storage Gen2 account to a new dataflow so you can use it in Power BI.
Learn more
Create and attach

Figure 9.6 – Creating dataflow types

For this recipe, we will connect to an external source, and we will use the **Define new tables** option by clicking on **Add new tables**. Other options will be explored in the rest of the chapter.

7. We are now creating a new dataflow from scratch, and we need to select a connector. A familiar section will appear from where you can select a data source to connect with (as you would do with **Get Data** in Power BI Desktop).

8. Select the **Web API** connector and enter the following URL to connect to a CSV file loaded in the Power Query Cookbook GitHub repository: `https://github.com/PacktPublishing/Power-Query-Cookbook/blob/main/Chapter09/FactInternetSales.csv`. Click on **Next** to see a data preview:

Figure 9.7 – Web API connector

A data preview page will appear where you can define, as you would in the desktop version, **File origin**, the **Delimiter** type, and **Data type detection**. Leave the options detected by default and click on **Transform data** to access the next section, the Power Query online UI:

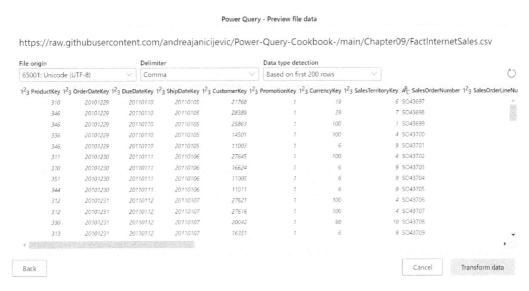

Figure 9.8 – Power Query data preview

The Power Query UI will pop up and from here, you can perform transformations and model your data as you would usually do, browsing different tabs and monitoring the applied steps on the right side of the UI:

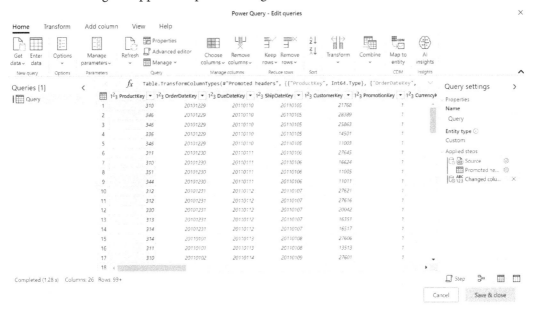

Figure 9.9 – Power Query online page

9. Rename the query and type `SalesData` under the **Query settings** section on the right side of the UI:

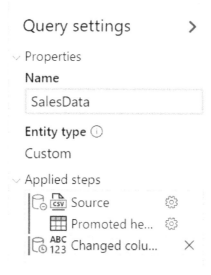

Figure 9.10 – Query settings

10. Click on the **Save & close** button on the bottom-right side to create the dataflow that contains this query:

Figure 9.11 – The Save & close button

11. The **Save your dataflow** window will pop up, where you can enter a name and short description for the dataflow. In this case, we will call the new dataflow `Sales` and then click on **Save**:

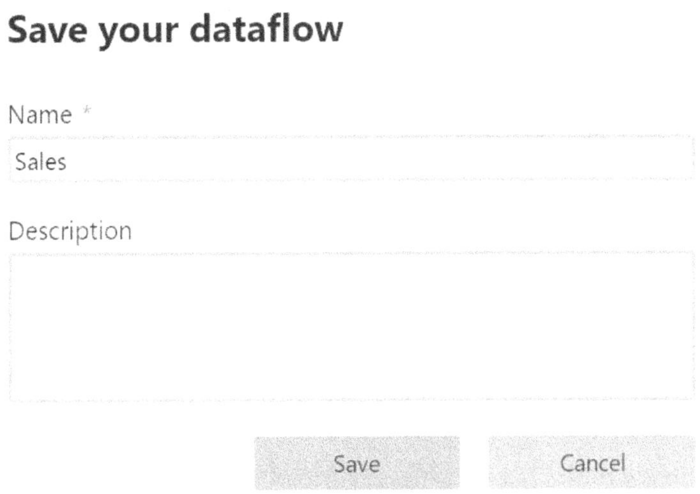

Save your dataflow

Name *

Sales

Description

Save   Cancel

Figure 9.12 – Save your dataflow

12. You can see in the following screenshot all the tables, called entities, within the dataflow that you created:

Figure 9.13 – Entity view within the dataflow

13. You could edit the dataflow, add other tables, and apply many other advanced features. In this case, we will close this view with the top-right button, **Close**:

Figure 9.14 – Close dataflow view

14. When closing the dataflow details view, you will go back to the workspace view and you will see the newly created content:

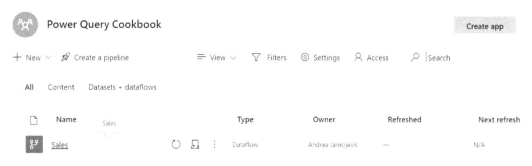

Figure 9.15 – Dataflows inside the workspace

15. Click now on the refresh icon to effectively load the data inside the storage of your Power BI Pro account:

Figure 9.16 – Refresh icon

You have seen how easy it is to create a dataflow. You can apply Power Query transformations and define data preparation steps directly from the browser and organize the content in workspaces.

Centralizing ETL with dataflows

Data preparation and transformation at an enterprise level, meaning managed centrally, compared to a self-service approach where you can perform your own data cleaning, is one of the most expensive and difficult tasks to manage within a company. Also called **centralized ETL**, this task is traditionally associated with enterprise tools, but with Power BI dataflows, you can **extract, transform, and load** data by connecting to data sources, transforming the data applying business logic, and then modeling the data to produce reports and do further analysis.

In this recipe, you will see how you can create multiple dataflows, and that these can be used by multiple users in their data models to produce reports and visualize all these pieces in the lineage view.

Getting ready

For this recipe, you need to have access to the Power BI portal, for which a Power BI Pro license is needed. You also need to have access to a Power BI workspace.

How to do it...

After you log in to the Power BI portal, perform the following steps:

1. Browse to your workspace, click on **New** and click on **Dataflow**.

2. Then click on **Add new tables** to connect to a data source.

3. Click on the **Web API** connector to connect to a CSV file from the GitHub repository.

4. Enter the following URL to connect to a CSV file loaded in the Power Query Cookbook GitHub repository: `https://github.com/PacktPublishing/Power-Query-Cookbook/blob/main/Chapter09/FactInternetSales.csv`. Click on **Next** to see a data preview.

5. A data preview page will appear, then click on **Transform data**.

6. Once the Power Query UI appears on the web page, rename the query to `InternetSales` under the **Query settings** section on the right side of the UI:

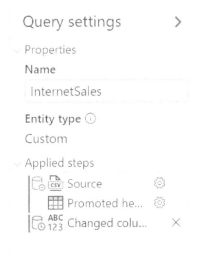

Figure 9.17 – Query settings

7. Let's apply some common Power Query transformation tasks such as **Choose columns**. Browse to the **Home** tab and click on the **Choose columns** button:

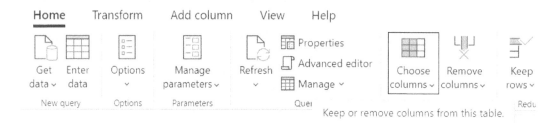

Figure 9.18 – Choose columns button

8. Select the ProductKey, OrderDateKey, OrderQuantity, TotalProductCost, SalesTerritoryKey, SalesAmount, and OrderDate columns and click on **OK**.

9. Now change the OrderData column's data type to **Date**:

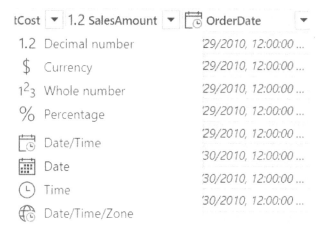

Figure 9.19 – Change the data type

10. Now let's save this query and click on the **Save & close** button at the bottom right of the UI:

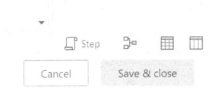

Figure 9.20 – Save & close button

11. Name this dataflow `Sales` and then click on **Save**.

12. Click on the **Close** button at the top right of the UI:

Figure 9.21 – Close dataflow view

13. Click on the refresh icon now to load the data in the underlying storage:

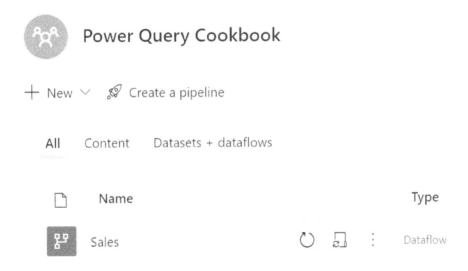

Figure 9.22 – Refresh icon

Following these steps, you have created the first dataflow. Now let's again perform *steps 1 to 6 and 10 to13* for the CSV files containing `DimProduct` and `DimTerritory`. The only steps not to perform are the data transformation ones from *7 to 9*.

When loading `DimProduct` and `DimTerritory`, refer to the following GitHub links when connecting through the Web API connector:

- `https://github.com/PacktPublishing/Power-Query-Cookbook/blob/main/Chapter09/DimProduct.csv`

- `https://github.com/PacktPublishing/Power-Query-Cookbook/blob/main/Chapter09/DimTerritory.csv`

When repeating *step 6*, rename the queries to `DimProduct` and `DimTerritory`, and at *step 11*, name the dataflows in the same way.

Once you have created these three dataflows, you will see them displayed within the workspace as in the following screenshot:

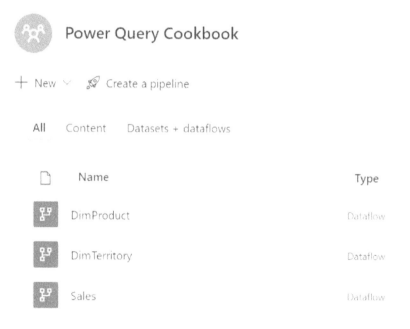

Figure 9.23 – Dataflows inside the workspace

Let's now see how we can use these three dataflows in Power BI. Once you open the Power BI Desktop application, perform the following steps:

1. Click on **Get data** and select the **Power BI dataflows** connector:

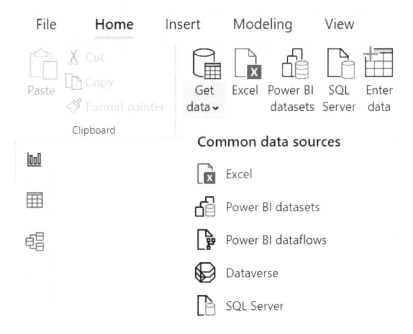

Figure 9.24 – The Power BI dataflows connector

2. Authenticate with your user account and click on **Connect**. In this way, you will be able to access workspaces you have permission to and connect to dataflows:

Figure 9.25 – Power BI dataflows authentication

3. The **Navigator** window will pop up, where you can see what dataflows you can connect to. Select the workspace where you have created the three dataflows `Sales`, `DimProduct`, and `DimTerritory`. In this case, select **Power Query Cookbook** and then expand the three dataflows' names and flag **DimProduct**, **DimTerritory**, and **InternetSales**. On the right side, you can see a preview of the data, as in the following screenshot:

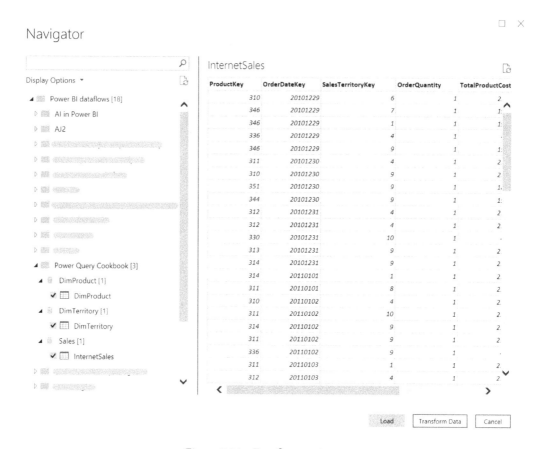

Figure 9.26 – Dataflow navigator

You can load and transform data as you normally do with any other data source. You can apply additional Power Query transformations from Power BI Desktop and then load data into your model. In this case, we will directly load the queries in the model.

4. After you have clicked **Load**, go to the **Model** view by clicking on the third icon tab on the left side of the UI:

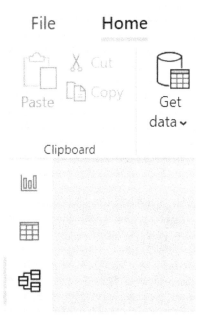

Figure 9.27 – Model view icon

Check the relationships between the `DimProduct` and `DimTerritory` tables and `InternetSales`. You should see the relationships that have already been detected as shown in *Figure 9.28* If not, hover your cursor on top of the `ProductKey` column in `DimProduct` and drag it to the same column in `InternetSales` and create a relationship. Repeat the same with `DimTerritory` to obtain something like in the following screenshot:

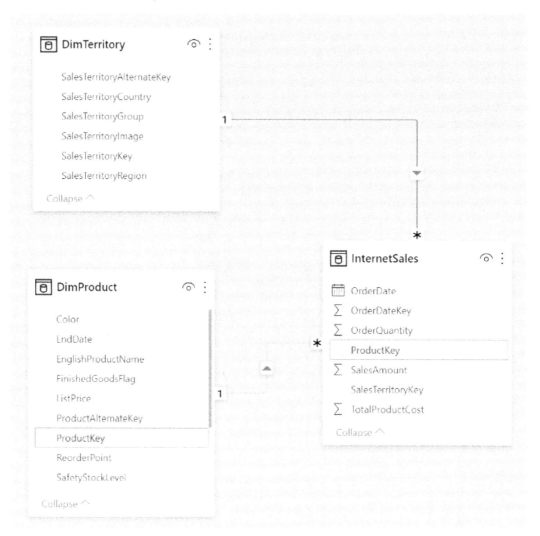

Figure 9.28 – Model view

5. Now go back to the **Report** view and create a visual as you like (in this recipe, we are focusing on the end-to-end flow; we will not focus on developing a report). In this case, we will quickly create a table like the following one:

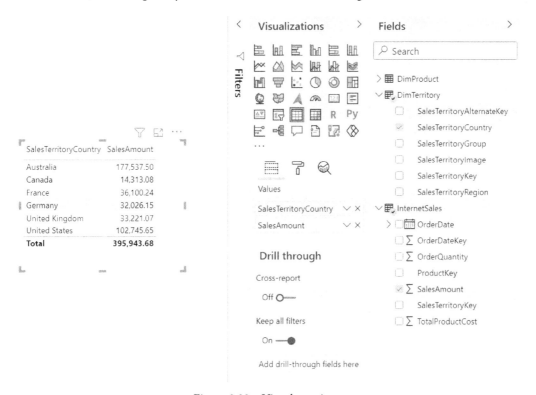

Figure 9.29 – Visual creation

6. Now let's publish the report and its dataset to a workspace you have access to (in this case, we are using the same workspace where we developed the dataflows, `Power Query Cookbook`). Save the `.pbix` file somewhere on your PC, browse to the **Home** tab and apply the **Sensitivity (preview)** label **General**, then click on the **Publish** button as in the following screenshot:

Figure 9.30 – Sensitivity labels

7. Select the workspace you want to publish to the report and its dataset, then click on **Select**:

Figure 9.31 – Publish to the workspace

You have now connected to your dataflows. You have created a Power BI dataset and a report on top of that dataset.

Now, again, open your browser and open the workspace where you have published these elements and follow the next steps:

1. Observe the elements displayed in the workspace:

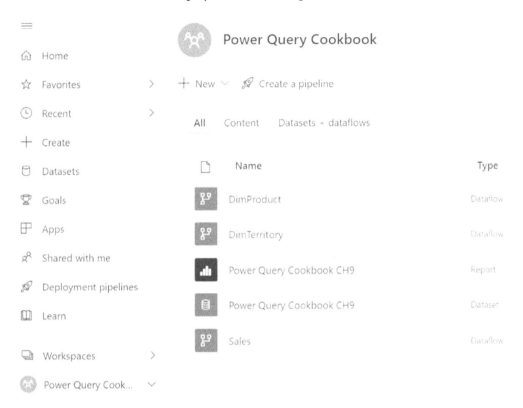

Figure 9.32 – Workspace view

2. Click on **View** and then on **Lineage** to see the preceding elements displayed in the lineage view:

Figure 9.33 – Lineage view button

3. See how the different artifacts (data sources, dataflows, the dataset, and the report) are displayed in a way that makes it easy for the user to understand how data moves:

Figure 9.34 – Lineage view

In general, instead of having datasets directly connected to the data sources, you can create an intermediate layer with the dataflows to which multiple users can connect to create their own data models and reports. In this way, you can create centralized ETL pipelines with a familiar low-code tool, Power Query, making them reusable and available to your organization. You do not need to necessarily connect to the original source data and then create multiple datasets but can use standardized dataflows, integrate them with external data, and create reports and dashboards on top of them.

Building dataflows with Power BI Premium capabilities

The Power BI service has different licensing options and the Premium one allows you to access some advanced capabilities with the dataflows feature. You can use computed entities and linked entities, which means you can link a new dataflow to an existing one without modifying the previous one. In this recipe, we will see these features in addition to the previous recipe, *Centralizing ETL with dataflows*.

Getting ready

For this recipe, you need to have access to Power BI Portal, for which you need a Power BI Pro license. You also need to have access to a Power BI workspace with Premium capacity. In this recipe, a Premium Per User license will be used.

Check this link to see the different licensing options:

```
https://docs.microsoft.com/en-us/power-bi/admin/service-admin-
licensing-organization#license-types-and-capabilities
```

How to do it...

For this recipe, you need to replicate the steps from the previous recipe to create the three dataflows in the workspace.

You should start this recipe by having a similar situation to this one:

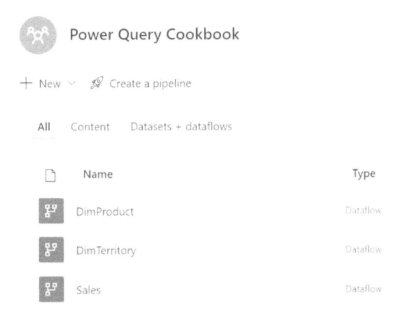

Figure 9.35 – Workspace view

Let's now follow the next steps to explore the Premium feature:

1. First, we need to assign a premium capacity to this workspace by clicking on **Settings**:

Figure 9.36 – Settings icon

2. The **Settings** window will appear on the right side of the screen. Click on the **Premium** tab and select a premium capacity option. In this case, **Premium per user** should be selected, and then click on **Save**:

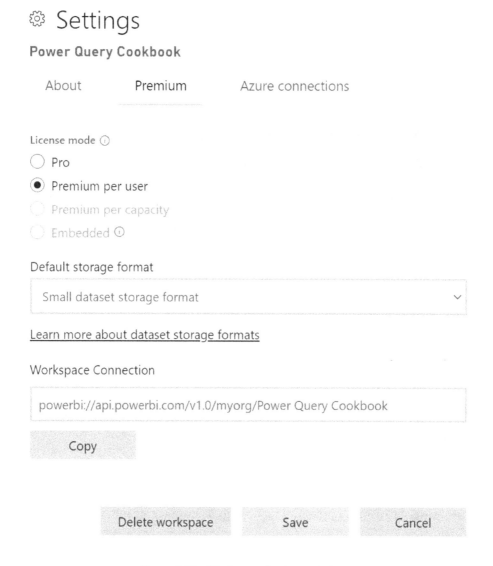

Figure 9.37 – Workspace Settings window

3. Check that the Premium Per User content is enabled by the presence of a diamond icon:

Figure 9.38 – Power BI Premium workspace

4. Now click on the **Sales** dataflow to see the tables contained in it:

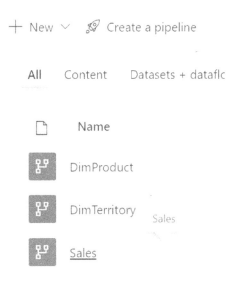

Figure 9.39 – Dataflow selection

5. You will see the previously loaded table, InternetSales, but now also want to add ResellerSales. Click on **Add tables** to open the connectors page:

Figure 9.40 – Add tables button

6. After having selected the **Web API** connector, enter the following URL to connect to a CSV file loaded in the Power Query Cookbook GitHub repository: https://github.com/PacktPublishing/Power-Query-Cookbook/blob/main/Chapter09/FactResellerSales.csv. Click on **Next** to see a data preview.

7. You will the data preview and click on **Transform data** to access the Power Query online page.

8. Rename the query to ResellerSales under **Query Settings**.

9. Click on the **Choose columns** button from the **Home** tab:

Figure 9.41 – Choose columns button

10. Select the ProductKey, OrderDateKey, OrderQuantity, TotalProductCost, SalesTerritoryKey, SalesAmount, and OrderDate columns and click on **OK**.

11. Then change the OrderDate column type by clicking on **Using locale…** since the date in this file is expressed in European format:

Figure 9.42 – Change the data type

12. Select **Date** for **Data type** and **English (United Kingdom)** for **Locale** and click on **OK** to correctly convert the Date column:

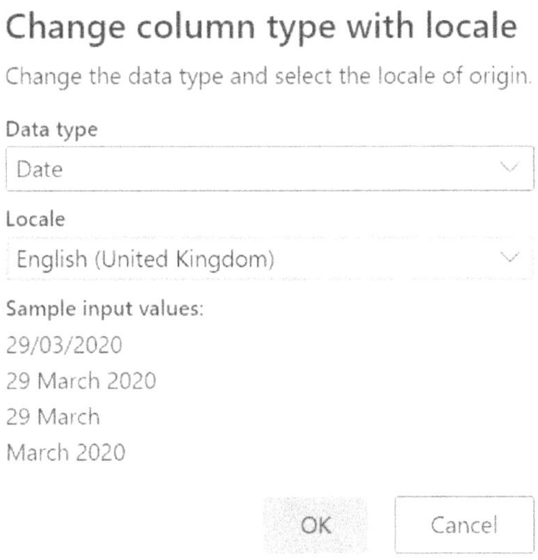

Change column type with locale

Change the data type and select the locale of origin.

Data type

Date   ∨

Locale

English (United Kingdom)   ∨

Sample input values:

29/03/2020
29 March 2020
29 March
March 2020

OK   Cancel

Figure 9.43 – Change the column type with locale

13. Now we want to append `InternetSales` and `ResellerSales` queries in one, but first, let's create a `Flag` column for each query to make the transactions identifiable as the `Internet` or `Reseller` channel. Select `ResellerSales` from the right side of the page, browse to the **Add column** tab, and select **Custom column**:

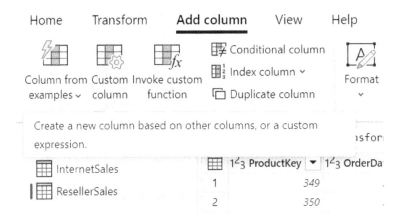

Figure 9.44 – Add a custom column

14. Create a column called `Channel`, write "`Reseller`", and click on **OK** to create this column with the value `Reseller` as a flag:

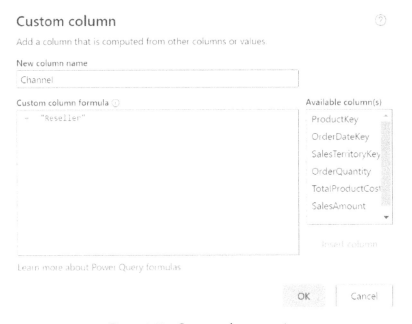

Figure 9.45 – Custom column creation

15. Repeat the previous two steps with `InternetSales` and write `Internet` instead of `Reseller`.

16. Now browse to the **Home** tab and click on **Combine**, on **Append queries,** and then on **Append queries as new** to append these two tables in to one:

Figure 9.46 – Append queries as new

17. The **Append** window will appear and from here, you can select **InternetSales** and **ResellerSales** as the two tables to be appended:

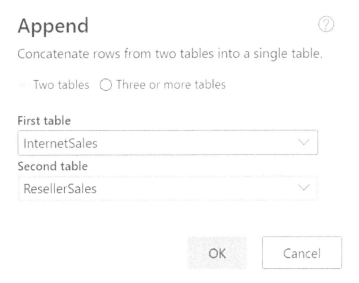

Figure 9.47 – Append window

18. The following message will appear and confirm that you are okay to have the data revealed by clicking on **Continue**:

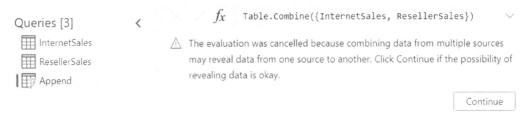

Figure 9.48 – Revealing data notification

19. You will see that a new query will be generated, called Append as a result of the transformation step. The lightning bolt on the table icon means that this is a **computed entity**:

Figure 9.49 – Computed entity icon

20. Rename the table to TotalSales under **Query Settings**.

21. Click on **Save & close** to save the transformations performed.

You have created a **computed entity**, which is an in-storage computation. The first two queries, InternetSales and ResellerSales, have been loaded in the underlying storage and when you perform additional transformations combining queries or referencing them, you are creating computed entities, in this case, TotalSales.

Now let's explore another concept through an example, **linked entities**. Imagine you want to combine some data in two tables that belong to different dataflows. You can do that by leveraging linked entities by following these steps:

1. From the workspace view, click on **New** and then on **Dataflow**:

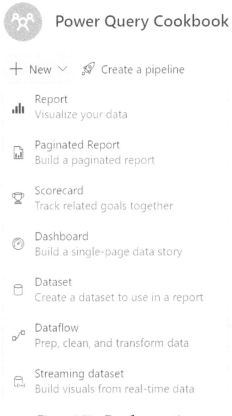

Figure 9.50 – Dataflow creation

2. Then click on **Add linked tables** to create a linked entity:

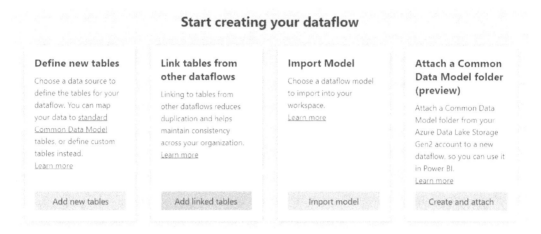

Figure 9.51 – Add linked tables

3. The Power BI dataflows connector page will pop up and you will be required to authenticate your account. Click on **Sign in**, complete the steps, and then click on the **Next** button, which will be activated after the authentication:

Figure 9.52 – Power BI dataflows connector

4. You will now see a preview of workspaces and dataflows you have access to. Expand the **DimTerritory** and **Sales** dataflows and select, respectively, the **DimTerritory** and **TotalSales** tables as in the following screenshot:

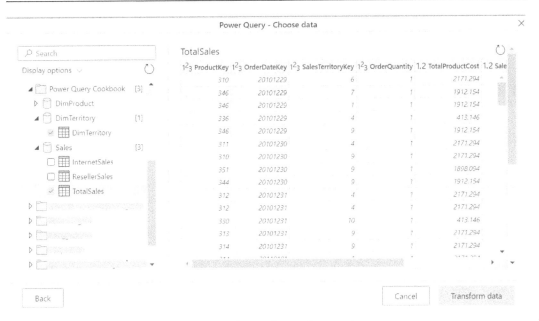

Figure 9.53 – Power Query data preview

5. You will see, under **Queries** on the left side of the page, the two tables, identified as linked entities by the chain icon. These tables are read-only, and you cannot perform any transformation directly on these tables:

Figure 9.54 – Linked tables icon

6. Since they are read-only tables, we will create a transformation and combine them in a third table. Select the **DimTerritory** query, browse to the **Home** tab, click on **Merge queries**, and then on **Merge queries as new** as in the following example:

Figure 9.55 – Merge queries as new

7.   Select **DimTerritory** for **Left table for merge**, **TotalSales** for **Right table for merge**, and the **SalesTerritoryKey** column for both queries. Then select **Left outer** under **Join kind** and click on **OK**:

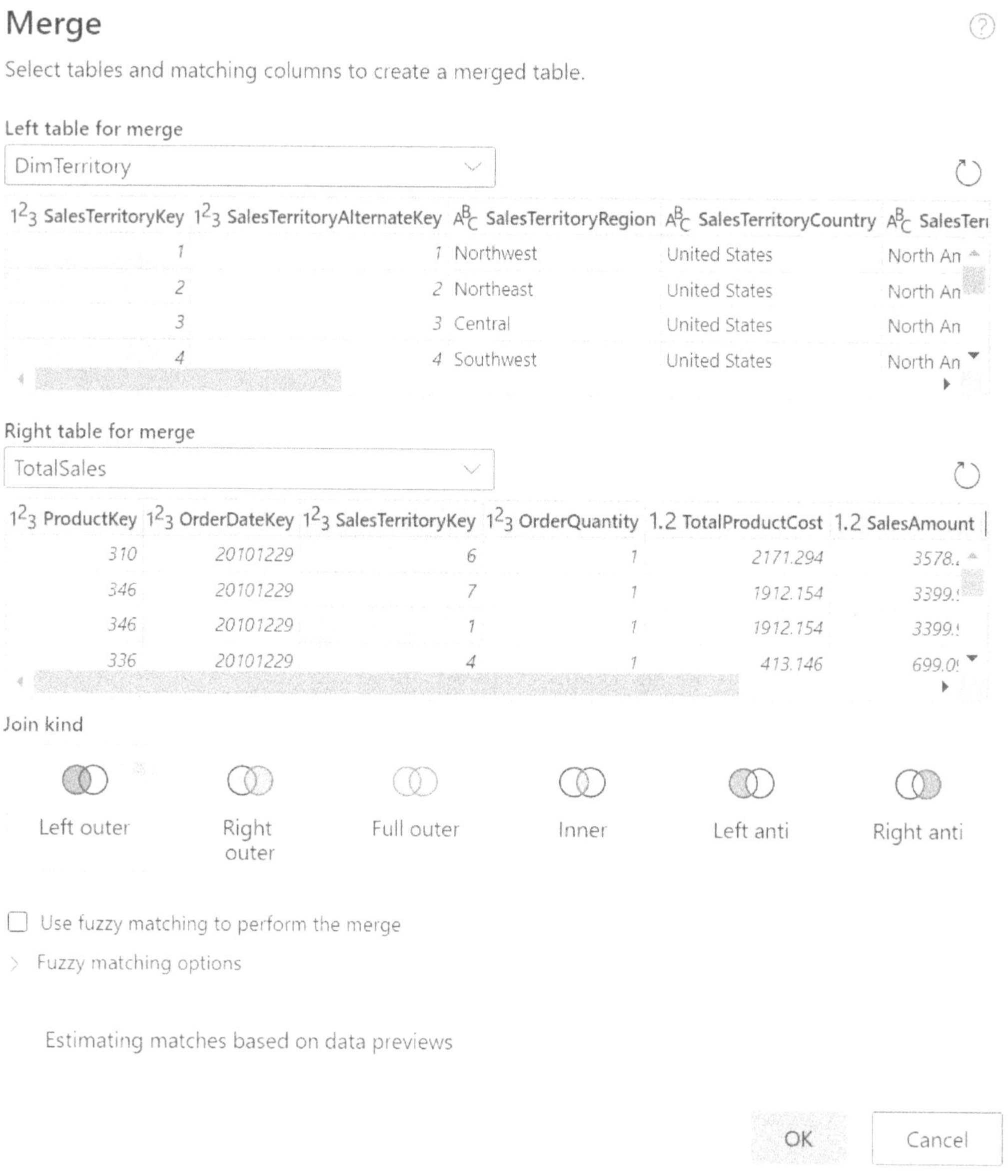

Merge   ⑦

Select tables and matching columns to create a merged table.

Left table for merge

| DimTerritory   ⌄ | ↻ |

1²₃ SalesTerritoryKey	1²₃ SalesTerritoryAlternateKey	A$_C$ SalesTerritoryRegion	A$_C$ SalesTerritoryCountry	A$_C$ SalesTerr
1		1 Northwest	United States	North An ▲
2		2 Northeast	United States	North An
3		3 Central	United States	North An
4		4 Southwest	United States	North An ▼

◀   ▶

Right table for merge

| TotalSales   ⌄ | ↻ |

1²₃ ProductKey	1²₃ OrderDateKey	1²₃ SalesTerritoryKey	1²₃ OrderQuantity	1.2 TotalProductCost	1.2 SalesAmount
310	20101229	6	1	2171.294	3578.. ▲
346	20101229	7	1	1912.154	3399.!
346	20101229	1	1	1912.154	3399.!
336	20101229	4	1	413.146	699.0! ▼

◀   ▶

Join kind

◎	◎	◎	◎	◎	◎
Left outer	Right outer	Full outer	Inner	Left anti	Right anti

☐ Use fuzzy matching to perform the merge

› Fuzzy matching options

Estimating matches based on data previews

OK   Cancel

Figure 9.56 – Merge window

8. A new query called **Merge** will be created and you can observe how the newly created table is a computed entity:

Figure 9.57 – Computed table icon

9. The new table contains **DimTerritory** columns and a column to expand with values coming from **TotalSales**. Click on the expand icon of the **TotalSales** column, flag **TotalProductCost** and **SalesAmount**, and then click on **OK**:

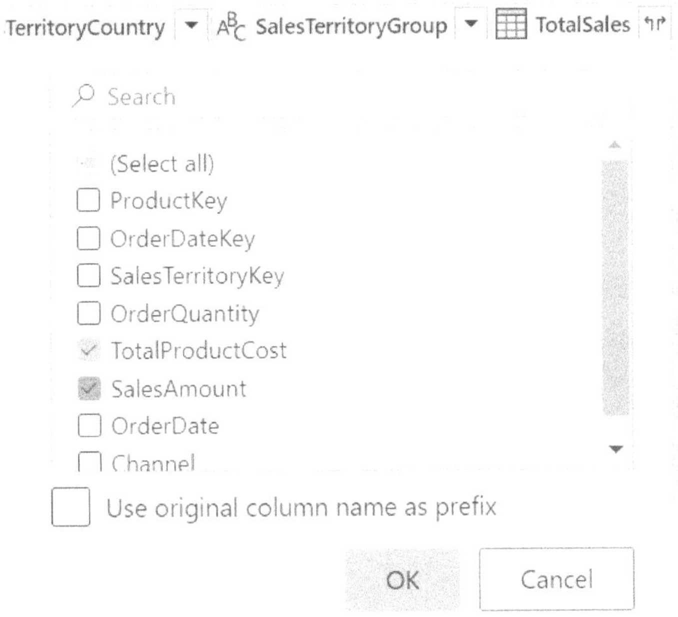

Figure 9.58 – Expand column

10. You have two new additional columns with sales data by geography:

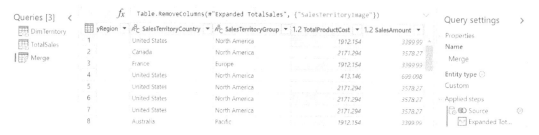

Figure 9.59 – Power Query page

11. Rename the query **Merge** to SalesGeo.

12. Click on **Save & close**.

13. Name the dataflow Sales Geography and click on **Save**.

14. Click on **Close** to go back to the workspace view:

Figure 9.60 – Close dataflow view

15. Now click on **View** and then on **Lineage** to access the lineage view:

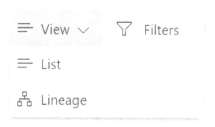

Figure 9.61 – Lineage button

16. You can see, in the lineage view, how the different dataflows depend on each other and from where data is coming:

Figure 9.62 – Lineage view

Linked entities and computed entities are very useful when it comes to minimizing data replication and maximizing data consistency. You can create dataflow derivatives without reconnecting to the data sources, just by referring to existing dataflows.

Understanding dataflow best practices

When you create dataflows and use computed and linked entities, there are some suggestions on how to best optimize the creation, the configuration, and the use of dataflows in the Power BI environment. In this recipe, we will see some common best practices to take into consideration when modeling data with this powerful tool:

- Organizing your dataflows
- Structuring a data refresh
- Understanding the Common Data Model and Azure Data Lake Storage integration

Getting ready

For this recipe, you need to refer to the outputs of the previous two recipes and follow the reasoning presented.

It will be easier to follow the recipe's structure after having already created some dataflows.

How to do it...

Power BI dataflows is a powerful tool for centralized data preparation with a low-code approach and to know how to best leverage it, it is important to follow some suggestions for tasks that have always been associated with IT departments, such as extract-transform-load and staging tables.

Organizing your dataflows

The first thing to point out is that you should not have everything in one dataflow and you should not have all dataflows within one workspace.

Since dataflows are meant to be reused by multiple users for different workloads, it could become unmanageable to have all tables in one or a few dataflows. It is important to generally create two types of dataflows:

- **Staging dataflows**: You can create these dataflows to make the first load of data coming from original data sources (databases and source systems) and create the first layer of data that does not have any type of transformation, like in the following example:

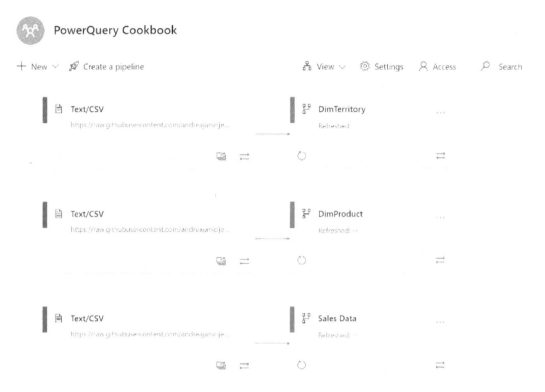

Figure 9.63 – Lineage view for staging dataflows

In this case, you have raw data inside this workspace to which you can grant permissions to users to access and create their transformations.

- **Transformation dataflows**: This is the second layer of data that references not the original data sources, but the first layer, the staging dataflows.

Users have permissions to create linked tables and compute entities and build their dataflows, all pointing to one single source of truth, the staging ones. If you check the recipes in this chapter, you will see how you can combine tables coming from different dataflows and obtain something more complex like the following example within the same workspace as we saw in the previous recipe.

You can combine tables coming from different dataflows and obtain something more complex from other workspaces as in the following example:

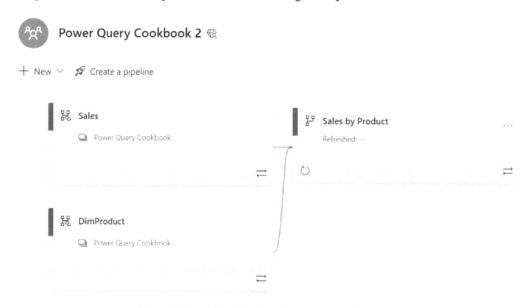

Figure 9.64 – Linked entities in separate workspaces

In this case, we are referencing two dataflows coming from the **Power Query Cookbook** workspace and a `Sales by Product` table was created in the **Power Query Cookbook 2** workspace by executing a computed entity.

You can create and manage multiple layers of dataflows and you can distribute them across different workspaces according to company departments and different project use cases.

If you see that a table is used frequently across many datasets, then you may find it useful to create a centralized dataflow and make it available to all users who need that table. Step by step, you can manage to create a complex and efficient extract-transform-load data architecture where you can control, thanks to the lineage view, where your data is most used with a low-code and business-friendly approach.

Structuring a data refresh

Another key topic regarding dataflows is refreshing data. Linking to the previous section, you may define a dataflow distribution across dataflows also according to refresh frequencies.

Tables with different refresh frequencies should stay in separate dataflows and, when possible, you should implement an incremental refresh.

You can set up an incremental refresh when you click on a dataflow from the workspace view. Then, under the **ACTIONS** section, you can click on the **Incremental refresh** icon – the last one of the four icons displayed, as shown in red in the following screenshot:

TABLE NAME	TABLE TYPE	ACTIONS
InternetSales	Custom	
TotalSales	Custom	

Figure 9.65 – Incremental refresh icon

Once you click on the icon, a section on the right side of the page will pop up where you can enable the incremental refresh. Choose the **DateTime** column from your table and define the data storage options:

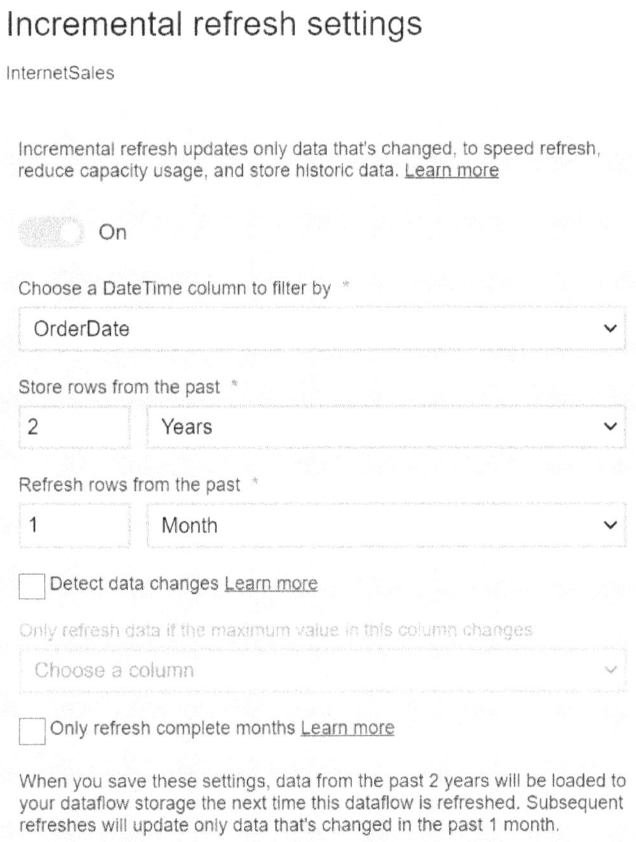

Incremental refresh settings

InternetSales

Incremental refresh updates only data that's changed, to speed refresh, reduce capacity usage, and store historic data. Learn more

On

Choose a DateTime column to filter by *

| OrderDate | ⌄ |

Store rows from the past *

| 2 | Years | ⌄ |

Refresh rows from the past *

| 1 | Month | ⌄ |

☐ Detect data changes Learn more

Only refresh data if the maximum value in this column changes

| Choose a column | ⌄ |

☐ Only refresh complete months Learn more

When you save these settings, data from the past 2 years will be loaded to your dataflow storage the next time this dataflow is refreshed. Subsequent refreshes will update only data that's changed in the past 1 month.

Figure 9.66 – Incremental refresh settings

You can check out *Leveraging on incremental refresh and folding* recipe in *Chapter 6, Optimizing Power Query Performance*, to see how to apply this feature to your data and how storage and data refresh logic works.

Let's now concentrate on the relationship between dataflows and datasets when it comes to refreshing.

Following the example from the *Centralizing ETL with dataflows* recipe, where we realized an end-to-end flow from source to report, it is important to point out that when you trigger the refresh for a dataflow, the dataset will not be refreshed. You must manage the refreshing of these two elements separately, at least from the UI:

Figure 9.67 – Lineage view for dataflows, datasets, and reports

You can also use external tools such as **Power Automate** or a **REST API** to trigger the refresh sequentially.

Understanding the Common Data Model and Azure Data Lake Storage integration

If you've already had the chance to read the previous recipes in this chapter, you will have seen that we mentioned the underlying Power BI storage where the data of dataflows is stored. You can access data stored in this underlying storage only with the Power BI dataflows connector and not from external tools.

This is something that happens by default if you do not define your storage resource to store data loaded in dataflows.

Power BI dataflows offer you the possibility to store data in Azure Data Lake Storage and access dataflow tables from external tools. You can create your data lake and link it to your Power BI environment in two ways:

- **At the organization level**, where all dataflows created will be stored inside one data lake. You can configure it from the **Admin portal** by clicking on **Azure connections**:

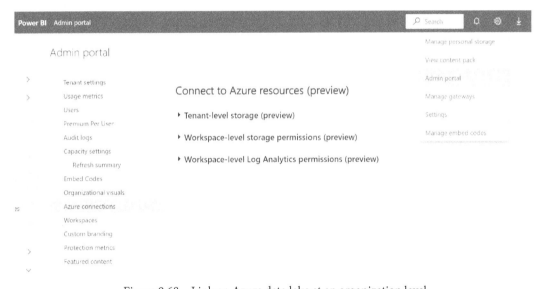

Figure 9.68 – Link an Azure data lake at an organization level

- **At the workspace level**, where dataflows within one workspace will be stored inside the specific data lake linked to that workspace. You can set it up from the workspace **Settings** and by clicking on **Azure connections**:

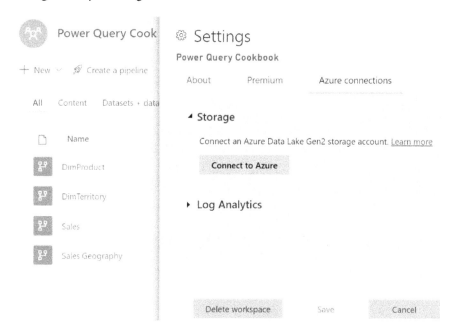

Figure 9.69 – Link an Azure data lake at the workspace level

Once you have linked your storage account, you can access this data from other tools within the Microsoft platform and also from third-party tools.

It is important to highlight that data is stored in a defined format, called the **Common Data Model** (**CDM**). The CDM is a shared data model, and it is a way of organizing data from different sources into a standard structure. The CDM includes over 340 standardized data schemas that Microsoft and its partners have published.

You can have a look at these schemas directly from the Power Query online UI. When you have a table open in edit mode, you can browse to the **Home** tab and click on **Map to entity**, which allows you to map your columns to a standard data structure:

Figure 9.70 – Map to entity button

You will open the **Map to CDM entity** window, where you can search entities, for example, **Product**, and then you can map the columns of your query output to attributes of a standard CDM entity. There are multiple entities, such as **Account**, **Address**, **Product**, **Social Activity**, and many more.

Map to CDM entity

Map the columns from your query output to attributes of a standard Common Data Model (CDM) entity. Learn more

Search	⋮≣ Auto map	⊗⋮ Clear all mappings			
PollSubmission	(none)	∨	packageContainer	Not mapped	
PortalLanguage	(none)	∨	parentProductId	Not mapped	
Position	(none)	∨	price	Not mapped	
PractitionerQualification					
PractitionerRole	(none)	∨	priceBase	Not mapped	
PriceList	(none)	∨	priceLevelId	Not mapped	
PriceListItem					
Procedure	(none)	∨	processId	Not mapped	
Product	(none)	∨	productId	Not mapped	
ProductAssociation	(none)		productNumber	Not mapped	
ProductRelationship	**Not mapped**		productStructure	Not mapped	
ProductSalesLiterature	A⁺c Channel		productStructure_display	Not mapped	
Property	▦ OrderDate				
PropertyAssociation	¹²₃ OrderDateKey		productTypeCode	Not mapped	
PropertyInstance	¹²₃ OrderQuantity		productTypeCode_display	Not mapped	
	¹²₃ ProductKey		productUrl	Not mapped	
Entity description	1.2 SalesAmount				
Information about products and their pricing information.	¹²₃ SalesTerritoryKey		quantityDecimal	Not mapped	
	1.2 TotalProductCost		quantityOnHand	Not mapped	
	(none)	∨	size	Not mapped	
	(none)	∨	stageId	Not mapped	

OK   Cancel

Figure 9.71 – Map to CDM entity window

The CDM allows the data to be unified in a well-known form with semantic meaning, and it simplifies the integration of data between different apps that use CDM standard entities, enabling these applications to easily read and understand the data.

But how is data organized in *Azure Data Lake Storage*? It is organized in folders and data is stored in CSV files. In the following example, you can see the backend structure of a dataflow saved in a workspace:

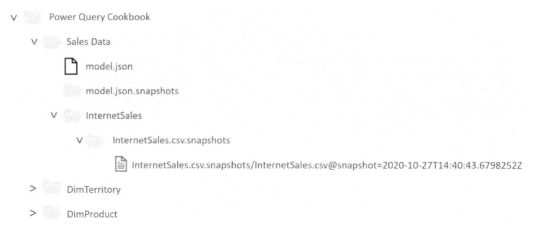

Figure 9.72 – CDM format in the Azure data lake

When you connect to Azure Data Lake Storage, when you load the first dataflow, a folder will be created underneath with the name of the workspace. Then a folder with the dataflow name is created and inside it, there are the following elements:

- `model.json`: In this file, you can find metadata, a set of data that describes and gives information about tables' column names, data types, mapping to standard entities, and more.

- `model.json.snapshots`: The same logic as the file before, storing historical metadata information from previous versions.

- `entity name`: In this case, `InternetSales`.

- **Entity name folder with CSV snapshots**: Historical CSV snapshots saved after each data refresh, in this case, `InternetSales.csv.snapshots`.

- `CSV file()`: The file where the current dataflow entity is stored, in this case, `InternetSales.CSV.snapshots/InternetSales.csv@snapshot=2020-10-27T14:40:43.6798252Z`.

If you add transformation steps and refresh your data, the above elements will be refreshed accordingly, and a snapshot of the previous version will be stored and accessible from external tools.

It is important to understand how CDM is structured because when you create a dataflow, you are offered other options on how to create dataflows, as shown in the following screenshot:

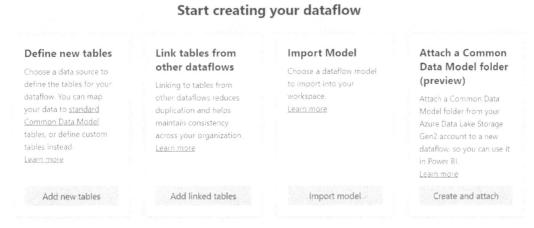

Figure 9.73 – CDM options to create dataflows

The other two options that you can use to leverage Power BI dataflows are the following:

- **Import Model**: If external systems generate the CDM format in the lake, you can attach that form here without connecting to source systems, but you can map in the Power BI workspace something that already exists in the lake.

- **Attach a Common Data Model folder (preview)**: You can attach an existing folder to the Azure data lake and make it visible from the Power BI service as a Power BI dataflow.

10

Implementing Query Diagnostics

Every transformation and data preparation step that you do has an impact on your performance when you develop Power Query transformations. You can analyze and test your queries to observe the consequences while refreshing your data and iteratively applying Power Query steps in Power BI Desktop.

In this chapter, we will see how to best use a built-in feature called **Query Diagnostics**, in order to easily retrieve relevant data on Power Query performance. In particular, we will go through the following recipes:

- Exploring diagnostics options
- Managing a diagnostics session
- Designing a report with diagnostics results
- Using Diagnose as a Power Query step

Technical requirements

For this chapter, you will be using the following:

- Power BI Desktop: `https://www.microsoft.com/en-us/download/details.aspx?id=58494`

The minimum requirements for installation are the following:

- .NET Framework 4.6 (Gateway release August 2019 and earlier)

- .NET Framework 4.7.2 (Gateway release September 2019 and later)

- A 64-bit version of Windows 8 or a 64-bit version of Windows Server 2012 R2 with current TLS 1.2 and cipher suites

- 4 GB disk space for performance monitoring logs

You can find the data resources referred to in this chapter at `https://github.com/PacktPublishing/Power-Query-Cookbook/tree/main/Chapter10`.

Exploring diagnostics options

You can use the Query Diagnostics tool to carry out an assessment of your queries and the steps performed. This means that once you have created Power Query steps, you can initiate a session that records and analyzes all the steps performed before you end that session. You can record what happens at the query level or deep dive at the single-step level. In this recipe, we will see where to find the Query Diagnostics option and how to set up the environment before starting a session.

Getting ready

For this recipe, you need to download the `FactInternetSales` CSV file.

In this example, we will refer to the `C:\Data` folder.

How to do it...

Once you open your Power BI Desktop application, you are ready to perform the following steps:

1. Click on **Get data** and select the **Text/CSV** connector.

2. Browse to your local folder where you downloaded the `FactInternetSales` CSV file and open it. The following window with a preview of the data will pop up. Click on **Transform Data**:

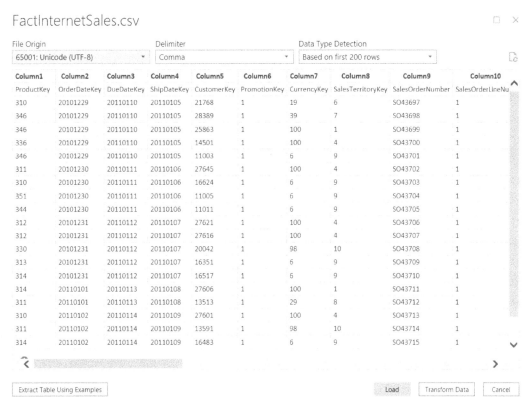

Figure 10.1 – Data preview

3. Browse to the **Tools** tab and you will see the different tools you can use to start a diagnostic session.

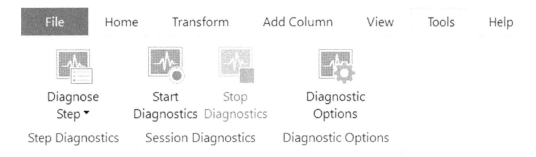

Figure 10.2 – Tools tab

You have the chance to diagnose a single step or start a general session and trace different actions, such as query refresh or the creation of a new step.

4. Before using this tool, we need to check the **Diagnostic Options** section. Click on the **Diagnostic Options** button, the last one in the **Tools** tab, and you will see the **Options** window pop up on the **Diagnostics** tab.

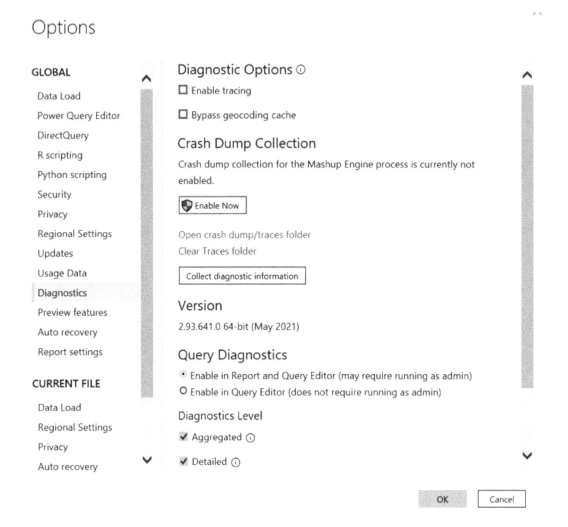

Figure 10.3 – Diagnostics tab in Options

5. Scroll down to the end and focus on the **Query Diagnostics** section, as shown in the following screenshot:

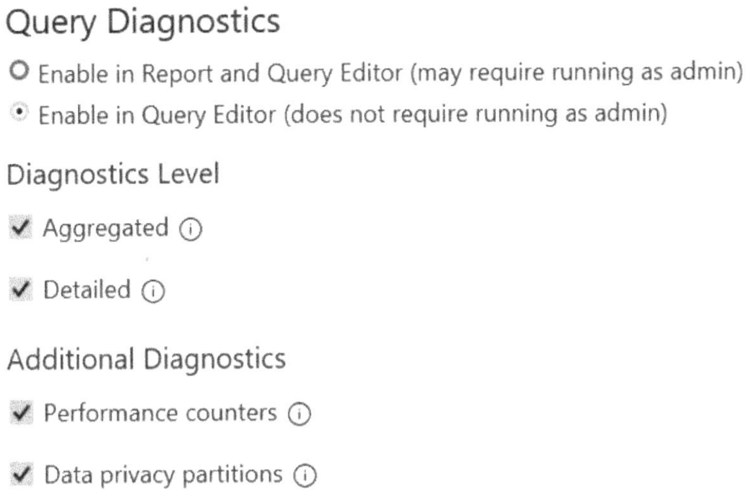

Query Diagnostics

○ Enable in Report and Query Editor (may require running as admin)

⦿ Enable in Query Editor (does not require running as admin)

Diagnostics Level

☑ Aggregated ⓘ

☑ Detailed ⓘ

Additional Diagnostics

☑ Performance counters ⓘ

☑ Data privacy partitions ⓘ

Figure 10.4 – Query Diagnostics in Options detail

You can enable diagnostics at the report and Query Editor levels, but this could require you to have admin rights. If you cannot run it, enable only **Query Editor**, as shown in the preceding screenshot.

You can select a diagnostics level and select which type of output to get:

*   **Aggregated**: You will easily understand diagnostics information because the information will be grouped allowing you to take immediate action.

*   **Detailed**: All diagnostics information is shown at the highest level of detail.

You will also have the chance to select an **Additional Diagnostics** option, such as **Performance counters** (including resource consumption and information about CPU and memory) and **Data privacy partitions** (logical partitions used to isolate steps for data privacy).

From here, you can define which type of information you want to see and analyze once you run the diagnostics session. Remember to check the information before using this tool.

Managing a diagnostics session

Once you set up the Query Diagnostics options as shown in the previous recipe, you can run a session and see what results you get thanks to this feature. In this recipe, we will perform some transformation steps, and then run a diagnostics session and observe the type of results.

Getting ready

For this recipe, you need to download the FactInternetSales CSV file.

In this example, we will refer to the C:\Data folder.

How to do it...

Once you open your Power BI Desktop application, you are ready to perform the following steps:

1. Click on **Get data** and select the **Text/CSV** connector.

2. Browse to your local folder where you downloaded the FactInternetSales CSV file and open it. A window with a preview of the data will pop up; click on **Transform Data**.

3. Browse to the **Home** tab and click on the **Choose Columns** button. The **Choose Columns** window will pop up. Flag the ProductKey, OrderDateKey, SalesTerritoryKey, OrderQuantity, ProductStandardCost, TotalProductCost, SalesAmount, and OrderDate columns and click on **OK**.

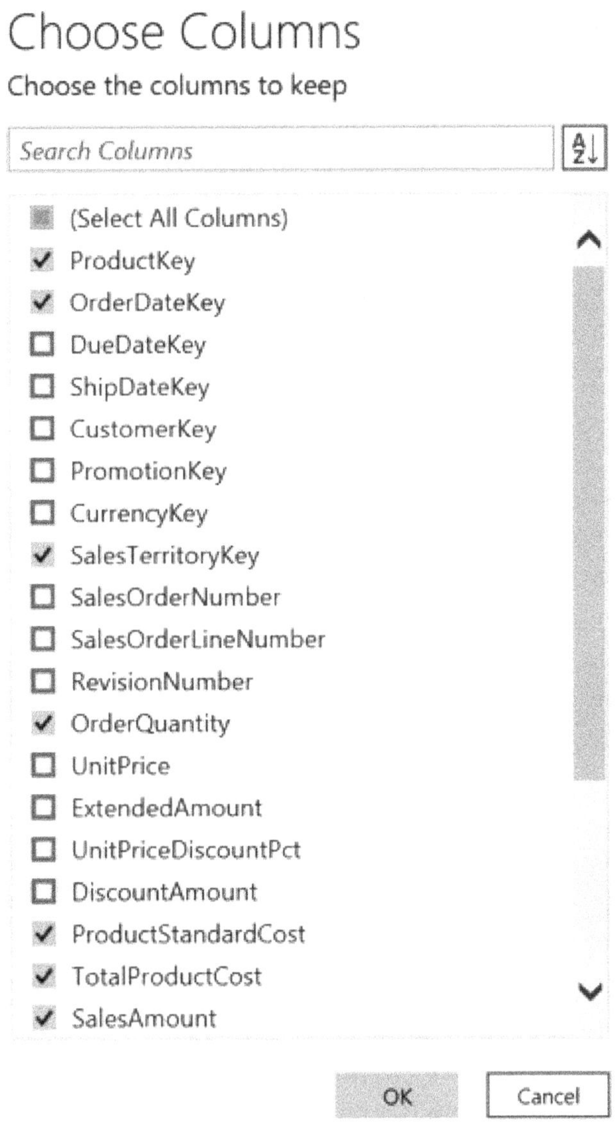

Choose Columns

Choose the columns to keep

| Search Columns | | A↓Z |

- ▨ (Select All Columns)
- ✔ ProductKey
- ✔ OrderDateKey
- ☐ DueDateKey
- ☐ ShipDateKey
- ☐ CustomerKey
- ☐ PromotionKey
- ☐ CurrencyKey
- ✔ SalesTerritoryKey
- ☐ SalesOrderNumber
- ☐ SalesOrderLineNumber
- ☐ RevisionNumber
- ✔ OrderQuantity
- ☐ UnitPrice
- ☐ ExtendedAmount
- ☐ UnitPriceDiscountPct
- ☐ DiscountAmount
- ✔ ProductStandardCost
- ✔ TotalProductCost
- ✔ SalesAmount

OK   Cancel

Figure 10.5 – Choose Columns

4. Change the `ProductKey` data type to **Text**.

Figure 10.6 – Changing the data type

5. Browse to the **Tools** tab and click on **Start Diagnostics**, as shown in the following screenshot:

Figure 10.7 – Start Diagnostics button

6. You will see that the **Start Diagnostics** icon will be deactivated, and **Stop Diagnostics** will be enabled. This means that the session is currently active, and it is recording all the steps you are doing.

   Now, browse to the **Home** tab and click on the **Refresh Preview** button to run the diagnostics on all Power Query steps performed previously at once.

Figure 10.8 – Refresh Preview button

7. After the refresh finishes, go back to the **Tools** tab and click on **Stop Diagnostics**.

Figure 10.9 – Stop Diagnostics button

8. Once you click the **Stop Diagnostics** button, the output of the session will be generated. Under the **Queries** section, you will find the three query outputs grouped in the **Diagnostics** folder, all created automatically by Power Query, as you can see in the following screenshot:

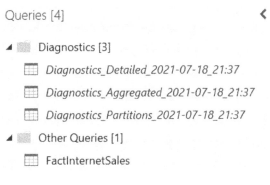

Figure 10.10 – Diagnostics output

9. If you click on the **Diagnostics_Detailed** query, you will see a query appear with data regarding the diagnostic session, as follows:

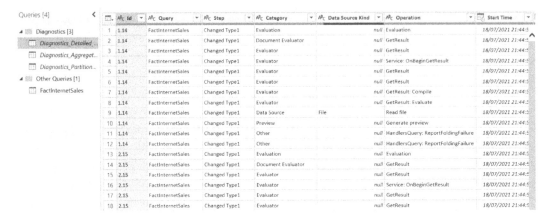

Figure 10.11 – Diagnostics schema

A high volume of information has been retrieved on the refresh we performed.

You will end up with three output queries, Diagnostics_Detailed, Diagnostics_Aggregated, and Diagnostics_Partitions, and they correspond to the output that you defined in the **Options** window in the *Exploring diagnostics options* recipe.

These three queries always have the same schema and you can read and interpret them to understand what has happened on the backend while Power Query was performing the refresh.

The most relevant dimensions can be summed up as follows:

- **Id**: Unique identifier for the evaluation of a single recording session.
- **Query**: Name of the query evaluated, listed under the **Queries** section on the left side of the UI.
- **Step**: Name of the applied step, listed under the **Query** settings pane on the right side of the UI.
- **Category**: The operation category.
- **Data Source Kind**: The data source you are accessing. In this example, it is **File**.

- **Operation**: The operation that is performed.

- **Start Time**: Operation start time.

- **End Time**: Operation end time.

- **Exclusive Duration (%)**: Time range of the event being active, expressed as a percentage.

- **Exclusive Duration**: The absolute time of the exclusive duration.

- **Resource**: Name of the resource you are accessing, in this case, the file path on your local PC.

- **Is User Query**: A true/false value that refers to whether the query was authored by the user (listed in the left-hand pane, **APPLIED STEPS**) or whether it was generated by some other user action.

- **Group ID**: Grouping created to approximate steps executed during the evaluation.

In the next recipe, we will see how we can design a report using this information.

Designing a report with diagnostics results

Once you have run the query diagnostics, it is important to know how to interpret these results. By just reading data from Power Query in a table, you can miss out on some relevant information. The best way to avoid this is to create a report on top of it by importing the diagnostics queries into a Power BI model and building visuals that can make sense out of that data.

Getting ready

For this recipe, you need to download the `FactInternetSales` CSV file.

In this example, we will refer to the `C:\Data` folder.

How to do it...

Once you open your Power BI Desktop application, you are ready to perform the following steps:

1. Click on **Get data** and select the **Text/CSV** connector.

2. Browse to your local folder where you downloaded the `FactInternetSales` CSV file and load it two times in order to make a comparison later in the recipe. You should have the following view:

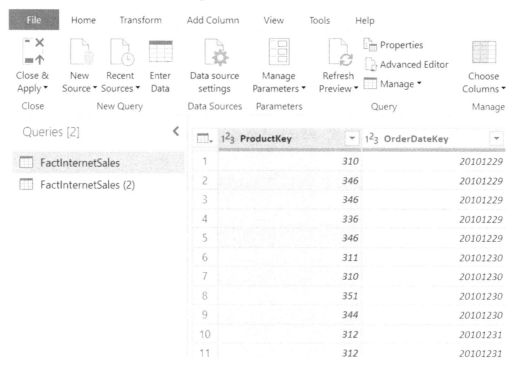

Figure 10.12 – Queries pane

3. Rename the `FactInternetSales` query to `FactInternetSales-example1` and `FactInternetSales (2)` to `FactInternetSales-example2`.

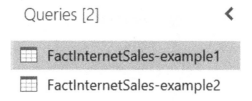

Figure 10.13 – Queries

4. Select `FactInternetSales-example1`, click on `OrderDateKey`, and then click on the **Remove Columns** button.

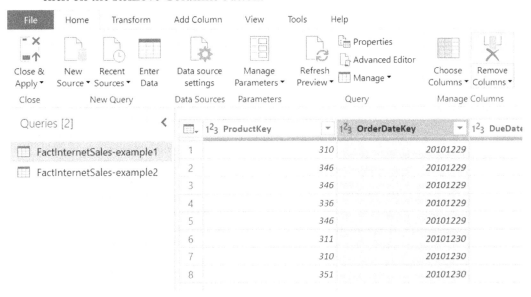

Figure 10.14 – Column selection

5. Now select the `ProductKey` column and change the type to **Text**.

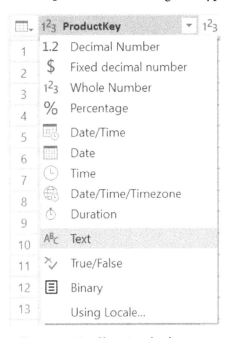

Figure 10.15 – Changing the data type

6. Select the `ShipDateKey` column and click on the **Remove Columns** button.

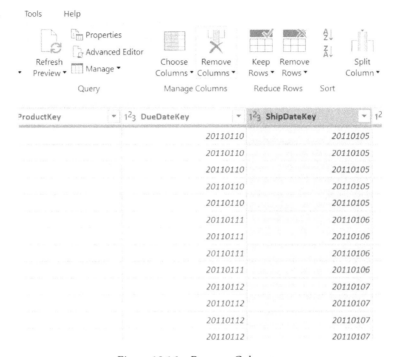

Figure 10.16 – Remove Columns

7. Now select `DueDateKey` and convert the data type to **Text** as you did in *Step 5* for `ProductKey`.

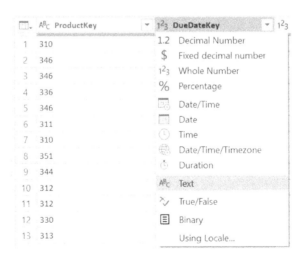

Figure 10.17 – Changing the data type

8. Now, browse to the end of the table, select the `OrderDate` column, and then click on the **Remove Columns** button to delete it.

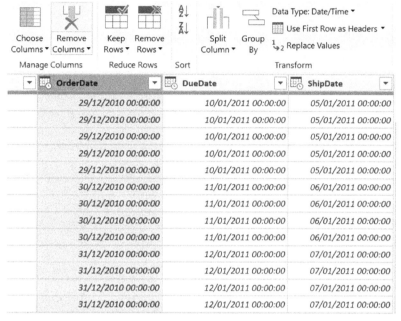

Figure 10.18 – Removing a column

9. Apply a filter to the `ProductKey` column by clicking on the drop-down icon, then **Text Filters**, and then **Begins With…**.

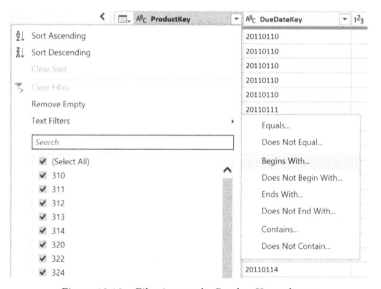

Figure 10.19 – Filtering on the ProductKey column

10. The **Filter Rows** window will pop up. Enter the value 3, as shown in the following screenshot, and click on **OK**:

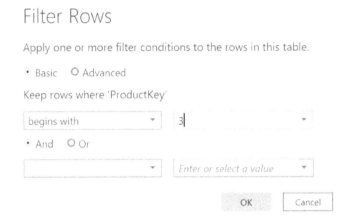

Figure 10.20 – Filter Rows window

11. You should have a situation like in the following screenshot under the **Query Settings** pane:

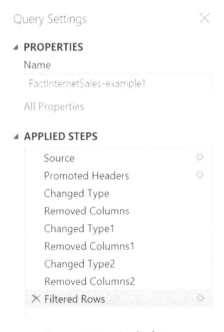

Figure 10.21 – Applied steps

We performed some steps, alternating between removing columns and data type changing steps.

This first part will help us make a comparison between two equal query outputs but with different step prioritizations that will be analyzed with Query Diagnostics.

Let's now work on the FactInternetSales-example2 query and replicate the next steps:

1. Select the FactInternetSales-example2 query and click on the **Choose Columns** button, as shown in the following screenshot:

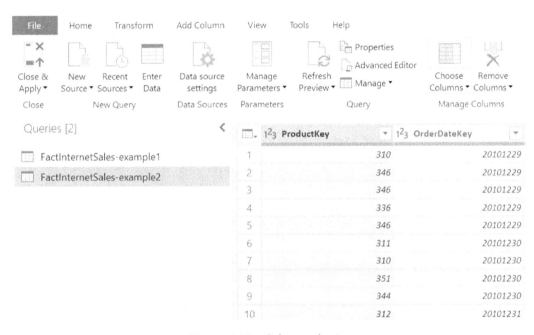

Figure 10.22 – Column selection

2. Remove the flag from the `OrderDateKey`, `ShipDateKey`, and `OrderDate` columns (the last one is at the end of the list) and click on **OK**.

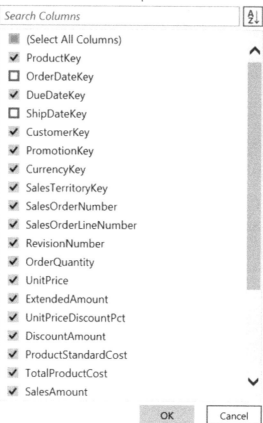

Figure 10.23 – Choose Columns

3. Select the `ProductKey` and `DueDateKey` columns, as shown in the following screenshot:

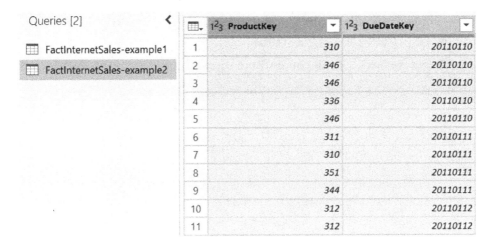

Figure 10.24 – Multiple-column selection

4. Right-click on one of the two columns and click on **Change Type** and then **Text** to convert the data type for both columns.

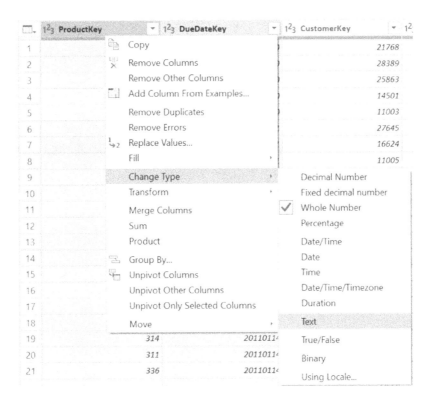

Figure 10.25 – Changing the data type for multiple columns

5. Apply a filter to the `ProductKey` column by clicking on the drop-down icon and then on **Text Filters** and **Begins With…**.

6. The **Filter Rows** window will pop up. Enter the value 3 and click on **OK**.

You should see the following steps for the second example:

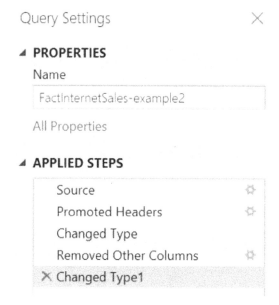

Figure 10.26 – Applied steps

As mentioned earlier, we now have two queries with the same output. Let's see how they perform and how much time it takes to apply the steps considering their different sequences.

Let's record a diagnostic session to see whether there are any differences between the two approaches by following the next steps:

1. Browse to the **Tools** tab and click on **Start Diagnostics**.

2. Then, browse to the **Home** tab and click on **Refresh Preview** and then **Refresh All**.

3. Go back to the **Tools** tab and click on **Stop Diagnostics**.

4. You should see, under the **Queries** pane on the left side of the UI, a folder named `Diagnostics` and four queries as diagnostics output.

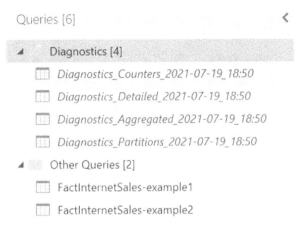

Figure 10.27 – Diagnostics output queries

5. Right-click on the query that starts with `Diagnostics_Aggregated` in order to analyze the key information that we need and click on **Enable load** to load the data in the final data model and build visualizations on top of it.

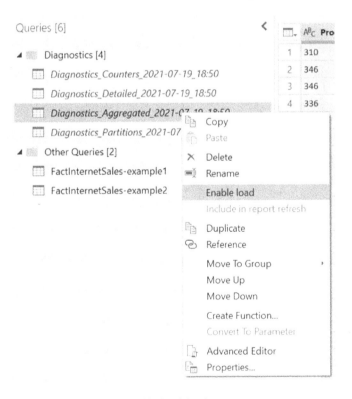

Figure 10.28 – Enable load for diagnostics queries

6. Click on the **Close & Apply** button to load the data and pass it to the visualization part.

Figure 10.29 – Close & Apply button

7. Under the **Fields** section in Power BI Desktop, you can see the diagnostics data. Select **Exclusive Duration** and **Query** under **Fields**, and then select **Stacked column chart**, as shown in the following example:

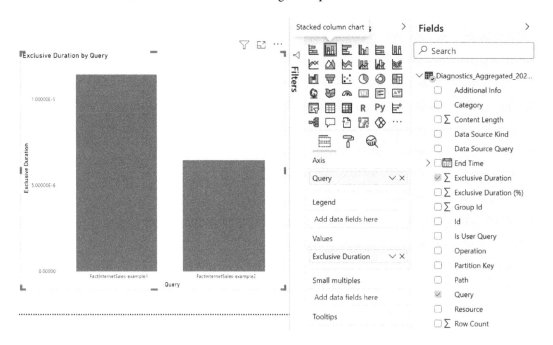

Figure 10.30 – Stacked column chart creation

You can see the exclusive duration of `FactInternetSales-example1` is almost double that of `FactInternetSales-example2`.

8.   Select **Category** for **Legend** to see the column chart broken down according to the **Category** type that was performed.

Figure 10.31 – Bar chart enrichment

You will see **Exclusive Duration by Query and Category** and you can spot what the operations taking more time to be applied are.

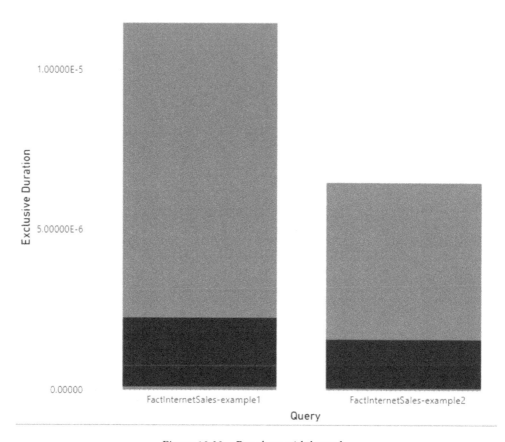

Figure 10.32 – Bar chart with legend

9. You can also create a table with the data you need. You can see the following example with a table with **Query**, **Start Time**, **Step**, **Exclusive Duration**, and **Category**:

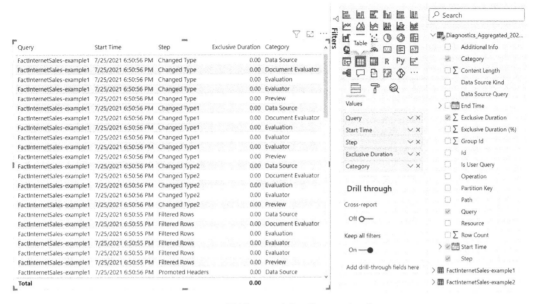

Figure 10.33 – Table visual for diagnostics data

10. In this example, the steps were performed in few-second fractions; in order to read them, we can add more decimal places for the values. Under the **Fields** section, click on **Exclusive Duration** to see the background behind the value, as in the following example:

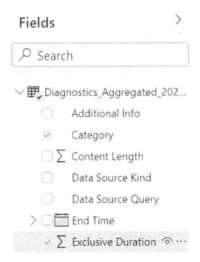

Figure 10.34 – Formatting a column by decimal type

11. After you have selected **Exclusive Duration**, the **Column tools** tab will pop up on the top section of the Power BI Desktop UI. Under that tab, in the **Formatting** section, enter the number of decimal places you want to see, in this case, 10, and press *Enter* on your keyboard.

Figure 10.35 – Column tools and Formatting

12. You will see how **Exclusive Duration** will now be easier to analyze on the table visual.

Query	Start Time	Step	Exclusive Duration	Category
FactInternetSales-example1	7/25/2021 6:50:56 PM	Changed Type	0.0000000149	Data Source
FactInternetSales-example1	7/25/2021 6:50:56 PM	Changed Type	0.0000000445	Document Evaluator
FactInternetSales-example1	7/25/2021 6:50:56 PM	Changed Type	0.0000000000	Evaluation
FactInternetSales-example1	7/25/2021 6:50:56 PM	Changed Type	0.0000001679	Evaluator
FactInternetSales-example1	7/25/2021 6:50:56 PM	Changed Type	0.0000011653	Preview
FactInternetSales-example1	7/25/2021 6:50:56 PM	Changed Type1	0.0000000105	Data Source
FactInternetSales-example1	7/25/2021 6:50:56 PM	Changed Type1	0.0000000463	Document Evaluator
FactInternetSales-example1	7/25/2021 6:50:56 PM	Changed Type1	0.0000000000	Evaluation
FactInternetSales-example1	7/25/2021 6:50:56 PM	Changed Type1	0.0000000306	Evaluator
FactInternetSales-example1	7/25/2021 6:50:56 PM	Changed Type1	0.0000003136	Evaluator
FactInternetSales-example1	7/25/2021 6:50:56 PM	Changed Type1	0.0000011197	Preview
FactInternetSales-example1	7/25/2021 6:50:56 PM	Changed Type2	0.0000000105	Data Source
FactInternetSales-example1	7/25/2021 6:50:56 PM	Changed Type2	0.0000000416	Document Evaluator
FactInternetSales-example1	7/25/2021 6:50:56 PM	Changed Type2	0.0000000000	Evaluation
FactInternetSales-example1	7/25/2021 6:50:56 PM	Changed Type2	0.0000001071	Evaluator
FactInternetSales-example1	7/25/2021 6:50:56 PM	Changed Type2	0.0000012885	Preview
FactInternetSales-example1	7/25/2021 6:50:55 PM	Filtered Rows	0.0000000080	Data Source
FactInternetSales-example1	7/25/2021 6:50:55 PM	Filtered Rows	0.0000002803	Document Evaluator
FactInternetSales-example1	7/25/2021 6:50:55 PM	Filtered Rows	0.0000000000	Evaluation
FactInternetSales-example1	7/25/2021 6:50:55 PM	Filtered Rows	0.0000000224	Evaluator
FactInternetSales-example1	7/25/2021 6:50:55 PM	Filtered Rows	0.0000001578	Evaluator
FactInternetSales-example1	7/25/2021 6:50:55 PM	Filtered Rows	0.0000009020	Preview
FactInternetSales-example1	7/25/2021 6:50:56 PM	Promoted Headers	0.0000000161	Data Source
Total			**0.0000179342**	

Figure 10.36 – Table visual

In general, you can see which values are contributing to the increase in duration of a query and how different factors can influence the evaluation of a certain query.

The most frequent analysis is related to what happens when the refresh from Power Query is started and to see what the impact of different steps could be. In this case, we can see how the overall duration is higher in `FactInternetSales-example1`, where we alternate removing columns and data type changes, whereas in the second, `FactInternetSales-example2`, we consolidate the steps, and it turns out to be, as we would expect, more performant.

You can leverage Query Diagnostics to make these comparisons and get the most out of this analysis.

There's more...

In general, as you had the chance to see with this recipe, it is better to consolidate the same steps, such as removing columns, filtering, and changing data types, and not alternate them like in the `FactInternetSales-example1` query.

Moreover, another key element that can be analyzed is query folding. As we saw in *Chapter 6*, *Optimizing Power Query Performance*, in the *Folding queries* recipe, you can send a query directly toward your data source and with Power Query Diagnostics, you can see what has been pushed back and review it. It is important to perform all transformations that support folding at the beginning to optimize performance.

Using Diagnose as a Power Query step

In the previous recipes, you had the chance to see how to run a query diagnostic at the query level, but you also have the chance to investigate single Power Query steps, without running general diagnostics for the entire query, and then drill down to the step you are interested in. In this recipe, we will see how to use this feature and test a single step.

Getting ready

For this recipe, you need to download the `FactInternetSales` CSV file.

In this example, we will refer to the `C:\Data` folder.

How to do it...

Once you open your Power BI Desktop application, you are ready to perform the following steps:

1. Click on **Get data** and select the **Text/CSV** connector.

2. Browse to your local folder where you downloaded the FactInternetSales CSV file and open it. A window with a preview of the data will pop up; click on **Transform Data**.

3. Select the ProductKey column, click on the drop-down icon, and click on **Number Filters** and then **Greater Than...**.

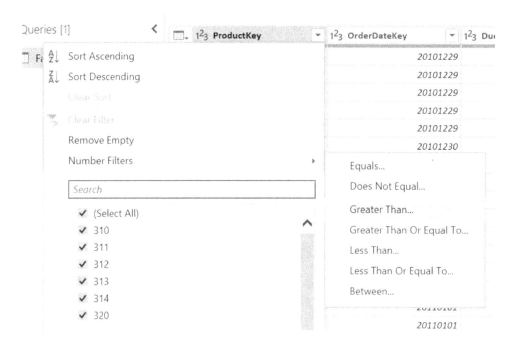

Figure 10.37 – Filtering on ProductKey

4. The **Filter Rows** window will appear. Enter the value 380 to keep rows where ProductKey is greater than that value and click on **OK**.

Filter Rows

Apply one or more filter conditions to the rows in this table.

⦿ Basic   ○ Advanced

Keep rows where 'ProductKey'

| is greater than ▾ | 380| ▾ |

⦿ And   ○ Or

| ▾ | *Enter or select a value* ▾ |

OK   Cancel

Figure 10.38 – Filter Rows window

5. We will now diagnose the step we performed in the previous step. Navigate to
    APPLIED STEPS on the right side of the Power Query UI, right-click on the
    Filtered Rows step, and click on **Diagnose**, as in the following screenshot:

Figure 10.39 – Diagnose step from APPLIED STEPS

You can also browse to the **Tools** tab and click on the **Diagnose Step** button to achieve the same result.

Figure 10.40 – Diagnose Step from the Tools tab

6. Under the **Queries** pane, you can see the output of Diagnose Step, similar to the ones seen in the previous recipes, but now, we focus only on that single step.

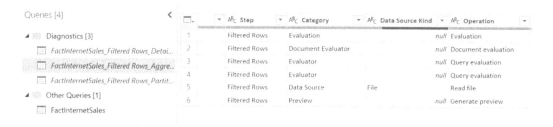

Figure 10.41 – Diagnose Step output

The diagnostic schema is the same as the one observed when we ran diagnostic queries in the previous sections of this chapter.

In general, Power Query offers you the possibility to evaluate the impact of different steps while developing your transformations and when refreshing queries in Power BI Desktop.

`Packt.com`

Subscribe to our online digital library for full access to over 7,000 books and videos, as well as industry leading tools to help you plan your personal development and advance your career. For more information, please visit our website.

Why subscribe?

- Spend less time learning and more time coding with practical eBooks and Videos from over 4,000 industry professionals

- Improve your learning with Skill Plans built especially for you

- Get a free eBook or video every month

- Fully searchable for easy access to vital information

- Copy and paste, print, and bookmark content

Did you know that Packt offers eBook versions of every book published, with PDF and ePub files available? You can upgrade to the eBook version at `packt.com` and as a print book customer, you are entitled to a discount on the eBook copy. Get in touch with us at `customercare@packtpub.com` for more details.

At `www.packt.com`, you can also read a collection of free technical articles, sign up for a range of free newsletters, and receive exclusive discounts and offers on Packt books and eBooks.

Other Books You May Enjoy

If you enjoyed this book, you may be interested in these other books by Packt:

Learn Power BI

Greg Deckler

ISBN: 978-1-83864-448-2

- Explore the different features of Power BI to create interactive dashboards
- Use the Query Editor to import and transform data
- Perform simple and complex DAX calculations to enhance analysis
- Discover business insights and tell a story with your data using Power BI
- Explore data and learn to manage datasets, dataflows, and data gateways
- Use workspaces to collaborate with others and publish your reports

Expert Data Modeling with Power BI

Soheil Bakhshi

ISBN: 978-1-80020-569-7

- Implement virtual tables and time intelligence functionalities in DAX to build a powerful model
- Identify Dimension and Fact tables and implement them in Power Query Editor
- Deal with advanced data preparation scenarios while building Star Schema
- Explore best practices for data preparation and data modeling
- Discover different hierarchies and their common pitfalls
- Understand complex data models and how to decrease the level of model complexity with different data modeling approaches

Packt is searching for authors like you

If you're interested in becoming an author for Packt, please visit `authors.packtpub.com` and apply today. We have worked with thousands of developers and tech professionals, just like you, to help them share their insight with the global tech community. You can make a general application, apply for a specific hot topic that we are recruiting an author for, or submit your own idea.

Share Your Thoughts

Now you've finished *Power Query Cookbook*, we'd love to hear your thoughts! Scan the QR code below to go straight to the Amazon review page for this book and share your feedback or leave a review on the site that you purchased it from.

`https://packt.link/r/1-800-56948-3`

Your review is important to us and the tech community and will help us make sure we're delivering excellent quality content.

Index

www.ingramcontent.com/pod-product-compliance
Lightning Source LLC
LaVergne TN
LVHW081329050326
832903LV00024B/1088